AMS SHORT COURSE LECTURE NOTES

Introductory Survey Lectures

published as a subseries of

Proceedings of Symposia in Applied Mathematics

Proceedings of Symposia in
APPLIED MATHEMATICS

Volume 59

An Introduction to Mathematical Modeling in Physiology, Cell Biology, and Immunology

American Mathematical Society
Short Course
January 8–9, 2001
New Orleans, Louisiana

James Sneyd
Editor

American Mathematical Society
Providence, Rhode Island

LECTURE NOTES PREPARED FOR THE
AMERICAN MATHEMATICAL SOCIETY SHORT COURSE
MATHEMATICAL BIOLOGY
HELD IN NEW ORLEANS, LOUISIANA
JANUARY 8–9, 2001

The AMS Short Course Series is sponsored by the Society's Program Committee for National Meetings. The series is under the direction of the Short Course Subcommittee of the Program Committee for National Meetings.

Cover art provided by James Sneyd. Used with permission.
A list of figure and table credits is included at the beginning of the volume.

2000 *Mathematics Subject Classification*. Primary 92C05, 92C20, 92C30, 92C37; Secondary 92D10, 92D30.

Library of Congress Cataloging-in-Publication Data

An introduction to mathematical modeling in physiology, cell biology, and immunology : American Mathematical Society, Short Course, January 8–9, 2001, New Orleans, Louisiana / James Sneyd, editor.

p. cm. — (Proceedings of symposia in applied mathematics, ISSN 0160-7634 ; v. 59. AMS short course lecture notes)

Includes bibliographical references and index.

ISBN 0-8218-2816-9 (alk. paper)

1. Physiology—Mathematical models—Congresses. 2. Cytology—Mathematical models—Congresses. 3. Immunology—Mathematical models—Congresses. I. Sneyd, James, II. American Mathematical Society. III. Proceedings of symposia in applied mathematics ; v. 59. IV. Proceedings of symposia in applied mathematics. AMS short course lecture notes.

QP33.6.M36 I585 2002
612′.001′5118—dc21   2002071734

∞ The paper used in this book is acid-free and falls within the guidelines
established to ensure permanence and durability.
Visit the AMS home page at `http://www.ams.org/`

10 9 8 7 6 5 4 3 2 1   07 06 05 04 03 02

Contents

Introduction

In the past few years there has been such a tremendous explosion of interest in mathematical biology that one could claim, without undue exaggeration, that biology is now one of the principal sources of mathematical applications. Spurred on in large part by advances in computing power, biological applications have reached deep into almost every traditional area of mathematics, and research in the generic field of mathematical biology is now so broad and vast that it has itself splintered into a multitude of separate sub-disciplines.

However, although mathematical biology has tremendous vitality and energy, there are still substantial barriers to any mathematician wishing to enter the field. Firstly, research in mathematical biology is done along very different lines than is most other mathematical research, as it is judged entirely on the quality of the science, not on the complexity or elegance of the mathematics involved. This requires not only a rather different mindset from the mathematician, but also from that mathematician's colleagues, who may be sitting on review or promotion committees. Secondly, it requires a substantial investment in time to learn the biological vocabulary and facts, and to establish collaborations with experimentalists. The personal skills necessary to work with a group of experimentalists, each with their own agenda and opinions, are not always those fostered by a degree in mathematics.

Breaking down these barriers is to the benefit of all mathematicians, as the close involvement of mathematics in the biological sciences greatly enriches both disciplines. In addition (to raise more mercenary points) such interdisciplinary efforts tend to be highly regarded by funding agencies and academic administrations.

The goal of this volume is to present a selected number of topics in mathematical biology to a mathematical audience. It aims to show how research in the field is done, what kind of mathematics is used, how one might best enter the field, what the outstanding questions are, as well as a brief historical survey of each topic so as to put current research into perspective. Because mathematical biology is such a huge field, ranging from studies of individual molecules such as DNA, to the study of entire populations, it is simply not possible to provide an overview of the entire field in a single volume. Thus, this volume consists of a series of talks covering a relatively restricted range of topics, with greater coverage of one topic, that of excitable cell physiology.

Electrically excitable cells form the basis of all neuronal activity and muscular contraction, and for this reason they have been a subject of intense investigation for well over a hundred years. Fortunately, because of the electrical nature of the cellular activity, it is possible to make experimental measurements with high accuracy and reproducibility, a fact that makes them ideal for mathematical investigation. Thus, it is in the study of electrically excitable cells that mathematics and physiology have traditionally had their closest links.

The opening chapter of this volume provides an introduction to the mathematics of electrically excitable cells. It discusses the basic theories of the action potential, including the Hodgkin-Huxley equations and the Morris-Lecar model. It provides the first glimpse of the complex types of oscillations found in various neurons, and of the complexities to be expected when individual neurons are connected to one another by synapses.

In the second chapter we delve into the visual system in more depth. This, again, involves the study of neurons, but this time in a different context from the previous chapter, and with a more specific physiological focus. This chapter discusses basic theories of edge detection, receptive fields, light adaptation, and orientation detection, including an overview of the structure of the visual system.

Muscle cells are a particular kind of electrically excitable cell, one designed to convert an electrical signal to a force. Thus their physiology is similar in many respects to that of other excitable cells, but the context is very different, and the mathematics has a different flavour. An understanding of the electrical properties of cardiac muscle is crucial for our understanding of what causes the heart to fail, an occurrence which kills millions of people every year. In the third chapter we discuss how we may use mathematics to study and understand cardiac arrhythmias.

The fourth chapter appears to digress somewhat from this overall theme of excitability, discussing as it does the dynamics of calcium inside cells which are, in general, not electrically excitable. This divergence, however, is less great than it might appear at first, as intracellular calcium homeostasis depends on the phenomenon of calcium excitability, as opposed to electrical excitability. The mathematical equations are similar to those of electrical excitability, but, once again, the context and physiology are quite different. The study of calcium dynamics is one of the most recent of the topics discussed in this volume, being only about eleven or twelve years old. It is a fine example of how new physiological observations can be understood, at least in part, by appealing to a more general mathematical theory that was developed in a different physiological context.

With those four chapters we leave the overview of the modelling of excitable cells, and their applications to physiological problems, and move on to a wider range of topics. From a very long list of possible topics we have chosen to discuss two that represent highly important areas of mathematical biology, areas in which there are already many mathematicians working.

The first topic, human genetics, involves modelling on a level rather more macroscopic than that of an individual cell, being concerned instead with the genetic properties of entire populations. In addition, the mathematics used is of a different kind, with much more of a probabalistic flavour. The particular topic presented, that of modeling the dynamics of genetic diseases in isolated populations, is an example of how analytical and numerical approaches can be combined to study large real-world data sets.

Finally, we discuss some models in immunology, particularly models of the HIV virus. For obvious reasons there is a tremendous amount of experimental and theoretical work being performed in this area. Our chapter highlights the (often overlooked) fact that very simple mathematical models can be used to great effect. As we show, even the simplest of mathematical models, when combined with experimental data, can be used to obtain results that, in hindsight, should have been obvious, but, without the model, rarely are.

Despite the breadth of this volume, the vast majority of mathematical biology remains unaddressed here. Our aim is merely to show how, in the few areas considered here, mathematics has played an important role in the study of a biological problem. In many respects, biology is the new frontier for applied mathematicians. We hope that the demonstration of how mathematics can be usefully used in biology will encourage yet more mathematicians to join our ranks.

James Sneyd
University of Auckland
New Zealand
August 2002

Figure and Table Credits

The American Mathematical Society gratefully acknowledges the kindness of the following institutions and individual in granting these permissions:

W. B. Saunders Co.

Figure 1, p. 58; from R. R. Rushmer, *Structure and function of the cardiovascular system*, 2nd ed., W. B. Saunders Co., Philadelphia, PA, 1976; with permission of W. B. Saunders Co.

Figure 2. p. 59; ibid; with permission of W. B. Saunders Co.

Figure 3, p. 61; ibid; with permission of W. B. Saunders Co.

Figure 12, p. 71; from D. P. Zipes and J. Jalife, *Cardiac electrophysiology; from cell to bedside*, 2nd ed., W. B. Saunders Co., Philadelphia, PA, 1995; with permission of W. B. Saunders Co.

Elsevier Science

Figure 13, p. 73; reprinted from A. V. Panfilov and J. P. Keener, *Reentry in 3-dimensional Fitzhugh-Nagumo medium with anisotropy*, Physica D **84** (1995), 545–552; with permission from Elsevier Science.

Springer-Verlag

Table 1, p. 89; from J. Keener and J. Sneyd, *Mathematical Physiology*, Springer-Verlag, New York, 1998; Table 5.1, p. 166; with permission from Springer-Verlag.

Table 3, p. 99; ibid., Table 12.1, p. 341; with permission from Springer-Verlag.

Figure 13, p. 112; ibid., Figure 12.10, p. 353; with permission from Springer-Verlag.

SIAM

Figure 1, p. 141; from A. S. Perelson and P. W. Nelson, *Mathematical models of HIV dynamics in vivo*, SIAM Review **41** (1999), 3–44, Figure 1.1, p. 5; courtesy of SIAM.

Figure 2, p. 142; ibid, Figure 1.2, p. 6; courtesy of SIAM.

Table 2, p. 155; ibid, Table 6.1, p. 30; courtesy of SIAM.

A. V. Panfilov

Figure 14, p. 74; with permission of A. V. Panfilov.

Proceedings of Symposia in Applied Mathematics
Volume **59**, 2002

Dynamics of Singularly Perturbed Neuronal Networks

David Terman

ABSTRACT. We consider models for neuronal systems and then discuss the sorts of activity patterns which may arise from the models. We first discuss models for single cells which display bursting oscillations. Several bursting classes are described. We then consider the dynamics of small networks of neurons and demonstrate how excitatory or inhibitory synaptic coupling may lead to either synchronous or asynchronous rhythms. Finally, we discuss larger networks. These networks are motivated by recent models for scene segmentation, thalamic rhythms and neuronal activity patterns in the basal ganglia.

1. Introduction

A fundamental goal of neuroscience is to understand how the nervous system communicates and processes information The basic structural unit of the nervous system is the individual neuron which conveys neuronal information through electrical and chemical signals. Patterns of neuronal signals underlie all activities of the brain. These activities include simple motor tasks such as walking and breathing and higher cognitive behaviors such as thinking, feeling and learning [**18, 20**].

Of course, neuronal systems can be extremely complicated. There are approximately 10^{12} neurons in the human brain. While most neurons consist of dendrites, a soma (or cell body), and an axon, there is an extraordinary diversity of distinct morphological and functional classes of neurons. Moreover, there are about 10^{15} synapses; these are where neurons communicate with one another. Hence, the number of synaptic connections made by a neuron can be very large; a mammalian motor neuron, for example, receives inputs from about 10^4 synapses.

An important goal of mathematical neuroscience is to develop and analyze mathematical models for neuronal activity patterns. The models are used to help understand how the activity patterns are generated and how the patterns change as parameters in the system are modulated. The models can also serve to interpret data, test hypotheses, and suggest new experiments. Since neuronal systems are typically so complicated, one must be careful to model the system at an appropriate level. The model must be complicated enough so that it includes those processes

---

1991 *Mathematics Subject Classification.* 92C20, 37N25, 34C15.

Key words and phrases. Neuronal Systems, Geometric Singular Perturbations, Oscillations, Synchronization.

The author was supported in part by NSF Grant #DMS9802339.

which are believed to play an important role in the generation of a particular activity pattern; however, it cannot be so complicated that it is impossible to analyze, either analytically or computationally.

A neuronal network's population rhythm results from interactions between three separate components: the intrinsic properties of individual neurons, the synaptic properties of coupling between neurons, and the architecture of coupling (i.e., which neurons communicate with each other). These components typically involve numerous parameters and multiple time scales. The synaptic coupling, for example, can be excitatory or inhibitory, and its possible turn on and turn off rates can vary widely. Neuronal systems may include several different types of cells as well as different types of coupling. An important and typically very challenging problem is to determine the role each component plays in shaping the emergent network behavior.

Models for the relevant neuronal networks often exhibit a rich structure of dynamic behavior. The behavior of even a single cell can be quite complicated. An individual cell may, for example, fire repetitive action potentials or bursts of action potentials that are separated by silent phases of near quiescent behavior [**17, 34**]. Examples of population rhythms include synchronized oscillations, in which every cell in the network fires at the same time and clustering, in which the entire population of cells breaks up into subpopulations or blocks; every cell within a single block fires synchronously and different blocks are desynchronized from each other [**13, 38**]. Of course, much more complicated population rhythms are possible. The activity may, for example, propagate through the network in a wave-like manner, or exhibit chaotic dynamics [**37, 54, 56**].

In this lecture, I will discuss models for neuronal systems and dynamical systems methods for analyzing these models. The discussion will focus primarily on models which include a small parameter and results in which geometric singular perturbation methods have been used to analyze the network behavior. I will not consider other types of models which are commonly used in the study of neural systems. The integrate and fire model of a single cell is one such example. A review of these types of models can be found in [**16, 22**].

An outline of the article is as follows. I will first present some of the basic biology used in modeling the neuronal systems. I will then discuss the explicit equations for the networks to be considered. Models for single cells are based on the Hodgkin-Huxley formalism [**15**] and the coupling between cells is meant to model chemical synapses. I will then consider models for single cells which exhibit bursting oscillations. There are, in fact, several different types of bursting oscillations, and there has been considerable effort in trying to classify the underlying mathematical mechanisms responsible for these oscillations [**17, 34**]. I will then discuss the dynamics of small networks of neurons. Conditions will be given for when these networks exhibit either synchronous or asynchronous rhythms. I will conclude by discussing an example of a larger network. These are referred to as *excitatory–inhibitory networks*; they display a rich structure of dynamic behavior and arise as models for several important neuronal systems. Some of the material in this article appeared in the review article [**38**].

2. A Little Biology

Here we discuss some very basic biological aspects of neurons and synapses. This will help to motivate the mathematical models which follow. For readers interested in learning more about the biology, some references are [**18, 20, 21**].

2.1. Single Cells.

The neuron is the basic information processing unit in the nervous system. Most neurons consist of a cell body (or soma) and a number of processes that project from the cell body; these are the dendrites and the axon. The dendrites spread out from the cell body in a tree-like manner. They collect incoming signals from other neurons or sensory receptors. Impulses are conducted away from the soma along the axon. Axons may be very short, while others may be very long (up to more than one meter). Many axons develop side branches called axon collaterals that help bring information to several parts of the nervous system simultaneously.

The neuron is surrounded by a cell membrane that maintains a stable resting potential between the outside and the inside of the cell. In response to a stimulus, the membrane potential may undergo a series of rapid changes, called an action potential. This results in the generation of a nerve impulse. In order to form a nerve impulse, the initial stimulus must be above some threshold amount. Properties of the nerve impulse, including its shape and propagation velocity, are often independent of the initial (superthreshold) stimulus.

The resting potential is created because there is an imbalance in the concentrations of certain ions between the inside and the outside of the cell. The intracellular concentrations of sodium and calcium ions are lower than in the extracellular space, while the extracellular potassium concentration is lower than inside the cell. In its resting state, the cell membrane is permeable to potassium; however it is virtually impermeable to sodium and calcium. The resting potential is about -70 mV and results, to a large extent, from the selective permeability of the membrane and the imbalance in concentration of potassium ions.

Following a stimulus, there is a sudden change in the permeability of the membrane to sodium (or calcium) ions. There are channels in the membrane selective to sodium. An action potential is generated when these channels open and sodium ions rush into the cell interior, causing a rapid rise in the membrane potential. After some delay, the sodium channels close and an increased number of potassium channels open. Potassium ions then flow to the outside of the cell and this brings the membrane towards the resting state. The membrane potential actually overshoots the resting potential. There is a recovery period in which sodium and potassium pumps move sodium and potassium ions, respectively, out of and into the cell until the resting membrane potential is achieved.

2.2. The Hodgkin-Huxley Equations.

The Hodgkin-Huxley equations were published in 1952 and describe the generation of action potentials in the squid giant axon [**15**]. However, the principles underlying the derivation of these equations form the basis of modeling other cells throughout the nervous system. The Hodgkin-Huxley model consists of four differential equations. One of these is a partial differential equation that describes the evolution of the membrane potential. The other three equations are ordinary differential equations which are related to properties of the ionic channels. The Hodgkin-Huxley equations can be written as:

$$
\begin{aligned}
C_M \frac{\partial v}{\partial t} &= D_M \frac{\partial^2 v}{\partial x^2} - g_{Na} m^3 h (v - v_{Na}) - g_K n^4 (v - v_K) - g_L (v - v_L) \\
\frac{\partial m}{\partial t} &= (m_\infty(v) - m)/\tau_m(v) \\
\frac{\partial n}{\partial t} &= (n_\infty(v) - n)/\tau_h(v) \\
\frac{\partial h}{\partial t} &= (h_\infty(v) - h)/\tau_n(v)
\end{aligned}
\tag{2.1}
$$

Here, $v(x,t)$ represents the membrane potential and each term in the first equation represents a separate current. Since the cell membrane separates charge it can be viewed as a capacitor and $C_M \frac{\partial v}{\partial t}$ is the capacitive current. The term $D_M \frac{\partial^2 v}{\partial x^2}$ represents longitudinal current along the axon and the remaining terms are ionic currents. The sodium current is $I_{Na} \equiv g_{Na} m^3 h (v - v_{Na})$. It is modeled using Ohm's law in which $g_{Na} m^3 h$ is the conductance and $(v - v_{Na})$ is a driving potential. The maximal sodium conductance is g_{Na} and $m^3 h$ can be thought of as the probability that a sodium channel is open. This will be discussed in more detail shortly. The constant v_{Na} is called the *sodium reversal potential.* This is the value of the membrane potential when the sodium concentration, which produces an inward flux of sodium through the sodium channel, is balanced by the electrical potential gradient tending to move sodium ions in the channel in the opposite direction. In a similar manner, $I_K \equiv g_K n^4 (v - v_K)$ is the potassium current with g_K being the maximal potassium conductance and n^4 is the probability that a potassium channel is open. Finally, $I_L \equiv g_L (v - v_L)$ is usually referred to as a leak conductance; it is due to the effects of other, less important, ions including chloride.

Note that the gating variables $m, h,$ and n satisfy differential equations that depend on the membrane potential v. Hence, the probability that a sodium or potassium channel is open or closed depends on the membrane potential. The sodium channel depends on two variables, namely m and h. We can think of the sodium channel as possessing two gates; both gates must be open in order for sodium to flow into the cell. We will not describe the voltage dependent steady-state functions $n_\infty(v), h_\infty(v), n_\infty(v)$ or the time constants $\tau_m(v), \tau_h(v), \tau_n(v)$ here. A detailed description can be found in [**18, 21**].

What distinguishes one cell from another are the types of ionic currents responsible for the action potential and what factors determine whether the channels corresponding to these ions are open or closed. In what follows, we simplify the presentation by ignoring the spatial dependence of the membrane potential; hence, we only consider systems of ordinary differential equations. Each cell then satisfies a system of the general form;

$$
\begin{aligned}
C_M v' &= -I_{ion}(v, w_1, w_2, ..., w_n) + I \\
w_i' &= \epsilon [w_{i,\infty}(v) - w_i]/\tau_i(v)
\end{aligned}
\tag{2.2}
$$

Here $v(t)$ denotes the membrane potential, I_{ion} is the sum of v- and t-dependent currents through the various ionic channel types and I represents external applied current. Each variable $w_i(t)$ describes the state of channels of a given type. Each current I_j is assumed to be ohmic and can be expressed as $I_j = \hat{g}_j\sigma_j(v, w_1, ..., w_n)(v - v_j)$ where $\hat{g}_j > 0$, the v_j are constants and the function σ_j represents the fraction of j-channels that are open. We will also consider models in which the concentration of intracellular calcium is a dependent variable and satisfies a single differential equation. More detailed description of the various terms in these equations can be found in [**35**].

2.3. Chemical Synapses. Synapses are the principal sites where neurons communicate with other neurons. There are two primary types of synapses. Most cells communicate with each other at chemical synapses in which messages are transmitted by the release of chemicals called neurotransmitters from the presynaptic cell and detected by receptors on the postsynaptic cell. There is a narrow gap or cleft between the pre- and postsynaptic membranes. Some cells transmit information to other cells by gap junctions in the cell membrane that form nonselective, low-resistant pores through which electrical current can flow. These are called electrical synapses. We will only consider models for chemical synapses; however, recently there has been an increased interest in gap junction coupling in the central nervous system. See, for example, [**19**].

Here we outline the sequence of events at a chemical synapse during neurotransmission. When an action potential reaches the presynaptic terminal, it activates voltage-gated calcium channels, allowing the entry of calcium ions. This triggers some unknown sequence of events leading to the fusion of vesicles containing neurotransmitters to the presynaptic membrane. The fusion of vesicles causes release of neurotransmitters that diffuse across the synaptic cleft and bind to specialized receptors in the postsynaptic membrane. This causes the opening of ion channels leading to a change in the membrane potential of the postsynaptic neuron. If this change in membrane potential exceeds some threshold amount, then the postsynaptic neuron may generate an action potential.

Note that the change in postsynaptic membrane potential may be either *depolarizing* or *hyperpolarizing*; that is, the membrane potential may increase or decrease. We say that the synapse is *excitatory* or *inhibitory* if the membrane potential is depolarizing or hyperpolarizing, respectively. At some synapses the binding of neurotransmitters to receptors do not directly gate ion channels. The activation of receptors may trigger secondary processes which then modulate ion channels. The indirect coupling of neurotransmitters to ion channels generally produces a slower response than those produced by the direct gating of channels by neurotransmitters.

3. An Example

3.1. Geometric Phase Space Approach. Throughout this section, we consider the system

$$
\begin{aligned}
v' &= -g_L(v - v_L) - g_K w(v - v_K) - g_{Ca} m_\infty(v)(v - v_{Ca}) + I \\
w' &= \epsilon(w_\infty(v) - w)\cosh((v + .1)/.3)
\end{aligned} \tag{3.1}
$$

These are the Morris-Lecar equations [**28**]; they can be viewed as a simple model for a neuron in which there are potassium and calcium currents along with the leak current. Here, v represents the membrane potential, rescaled between -1 and 1, and w is activation of the potassium current. Note that the activation of the calcium current is instantaneous. The small, positive parameter ϵ is introduced to emphasize that w evolves on a slow time scale. We treat ϵ as a small singular perturbation parameter in the analysis.

The parameters and nonlinear functions in (3.1) are given by $v_L = -.1$, $g_L = .5$, $g_K = 2$, $v_K = -.7$, $g_{Ca} = 1$, $v_{Ca} = 1$, $\epsilon = .1$, $m_\infty(v) = .5(1 + \tanh((v - .01)/.145))$, $w_\infty(v) = .5(1 + \tanh((v + .1)/.15))$.

If $I = 0$, then every solution approaches a stable fixed point. This corresponds to a neuron in its resting state. However, if $I = .4$, then solutions of (3.1) quickly approach a stable limit cycle. The periodic solution alternates between a silent phase and an active phase of near resting behavior. Moreover, there are sharp transitions between the silent and active phases.

These solutions can be easily analyzed using phase space methods. The v-nullcline is a cubic-shaped curve, while the w-nullcline is a monotone increasing function that intersects the v-nullcline at precisely one point; hence, there exists precisely one fixed point of (3.1), denoted by p_0. If $I = 0$, then p_0 lies along the left branch of the cubic v-nullcline and one can easily show that p_0 is asymptotically stable. If, on the other hand, $I = .4$, then p_0 lies along the middle branch of the cubic nullcline and p_0 is unstable. One can prove in this case that there must exist a stable limit cycle for all ϵ sufficiently small; moreover, the limit cycle approaches a singular limit cycle as shown in Figure 1 as $\epsilon \to 0$. Here $\{f(v, w) = 0\}$ and $\{g(v, w) = 0\}$ denote the $v-$ and $w-$ nullclines, respectively.

The singular limit cycle consists of four pieces. Two of these pieces lie along the left and right branches of the cubic nullcline; these correspond to the silent and active phases, respectively. The other two pieces correspond to the jumping up and down between the silent and active phases. We note that these pieces begin at one of the "knees" of the cubic nullcline.

The geometric analysis of (3.1) does not really depend on the the precise details of the nonlinear functions and other parameters. We used only that the v-nullcline is a cubic-shaped curve and the w-nullcline is a monotone increasing curve that intersects the cubic at a single fixed point. We also needed that $v' > 0$ (< 0) below (above) the cubic nullcline and that $w' < 0$ (> 0) above (below) the w-nullcline. In the more complicated examples that follow, we make similar geometric assumptions on the nonlinearities instead of specifying precisely defined functions and parameter values. The geometric assumptions can often be verified for a specific system using numerical methods.

3.2. Response to Applied Current. We now discuss how the geometric singular perturbation approach can be used to understand the response of a neuron to injected current. Consider the system

$$\begin{aligned} v' &= f(v, w) + I(t) \\ w' &= \epsilon g(v, w) \end{aligned} \tag{3.2}$$

where f and g denote the right hand sides of (3.1) with $I \equiv 0$ and $I(t)$ represents the injected current. We assume that when $I(t) = 0$ the system is excitableex-citable!system; that is, the v- and w-nullclines intersect along the left branch of the cubic. This is a globally stable fixed point and the model neuron is in its resting state. We further assume that there exists I_0 and $T_{on} < T_{off}$ such that

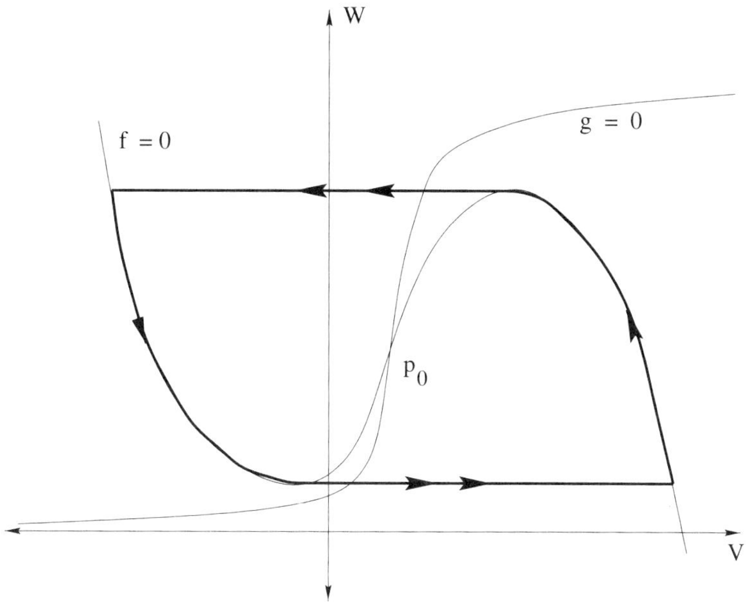

FIGURE 1. Nullclines and singular periodic orbit for an oscillatory relaxation oscillator.

$$I(t) = I_0 \text{ if } T_{on} < t < T_{off} \text{ and } I(t) = 0 \text{ otherwise.}$$

We will consider two cases: either $I_0 > 0$, in which case the injected current is said to be *depolarizing*, or $I_0 < 0$ and the injected current is *hyperpolarizing*. Figure 2 illustrates the neuron's response when (top) $I_0 = .1$ and (bottom) $I_0 = -.1$. In the depolarizing case, the neuron fires an action potential immediately after the injected current is turned on. The cell then returns to rest. In the hyperpolarizing case, the neuron's membrane potential approaches a more negative steady state until the current is turned off, at which time the neuron fires a single action potential. This last response is often called *post-inhibitory rebound* [**12**].

The geometric approach is very useful in understanding these responses. As before, we construct singular solutions in which ϵ is formally set equal to zero. See Figure 3. The singular solutions lie on the left or right branch of some cubic-shaped nullcline during the silent and active phases. The cubics depend on the values of $I(t)$. We denote the cubic corresponding to $I = 0$ as C and the cubic corresponding to I_0 as C_0. Note that if $I_0 > 0$, then C_0 lies 'above' C, while if $I_0 < 0$, then C_0 lies 'below' C. This is because of our assumption that $f < 0$ (> 0) above (below) the v-nullcline.

Consider the depolarizing case $I_0 > 0$. This is illustrated in Figure 3a. For $t < T_{on}$, the solution lies at the fixed point p_0 along the left branch of C. When $t = T_{on}$, $I(t)$ jumps to I_0 and the cell's cubic switches to C_0. If the left knee of C_0 lies above p_0 then the cell jumps up to the right branch of C_0; this corresponds to the firing of an action potential. If the w-nullcline intersects C_0 along its left branch, then the cell will approach the stable fixed point along the left branch of C_0 until the input is turned off. It is possible that the w-nullcline intersects C_0

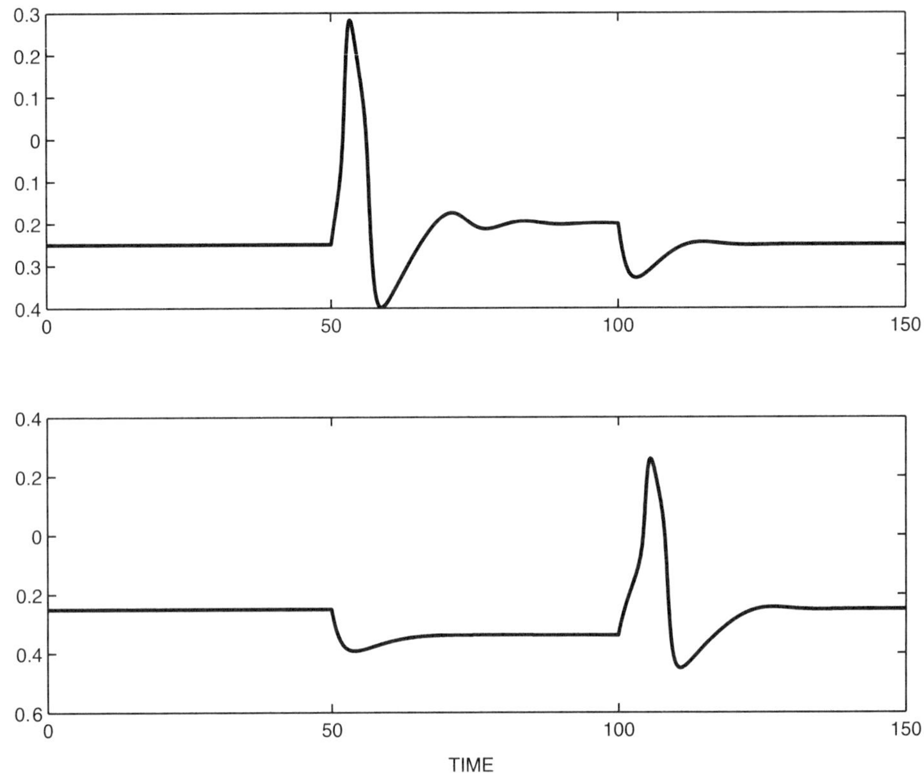

FIGURE 2. Response of a model neuron to applied current. Current is applied at time $t = 50$ and turned off at $t = 100$. In the top figure, the current is depolarizing ($I_0 = .1$), while in the bottom figure the current is hyperpolarizing ($I_0 = -.1$) and the neuron exhibits post-inhibitory rebound.

along its middle branch. If this is the case then the cell oscillates, continuing to fire action potentials, until $t = T_{off}$ when the input is turned off. Note that in order for the cell to fire an action potential, the injected current must be sufficiently strong. I_0 must be large enough so that the p_0 lies below the left knee of C_0.

We next consider the hyperpolarizing case $I_0 < 0$, shown in Figure 3b. Then C_0 lies below C and the w-nullcline intersects C_0 at a point denoted by p_1. When $t = T_{on}$, the solution jumps to the left branch of C_0 and then evolves along this branch approaching p_1 for $T_{on} < t < T_{off}$. When $t = T_{off}$, I switches back to 0 and the cell now seeks the left or right branch of C. If, at this time, the cell lies below the left knee of C, then the cell will jump up to the active phase giving rise to post-inhibitory rebound. In order to have post-inhibitory rebound, the hyperpolarizing input must be sufficiently large and last sufficiently long. I_0 must be sufficiently negative so that p_1 lies below the left knee of C. Moreover, $T_{off} - T_{on}$ must be sufficiently large so that the cell has enough time to evolve along the left branch of C_0 so that it lies below the left knee of C when the input is turned off.

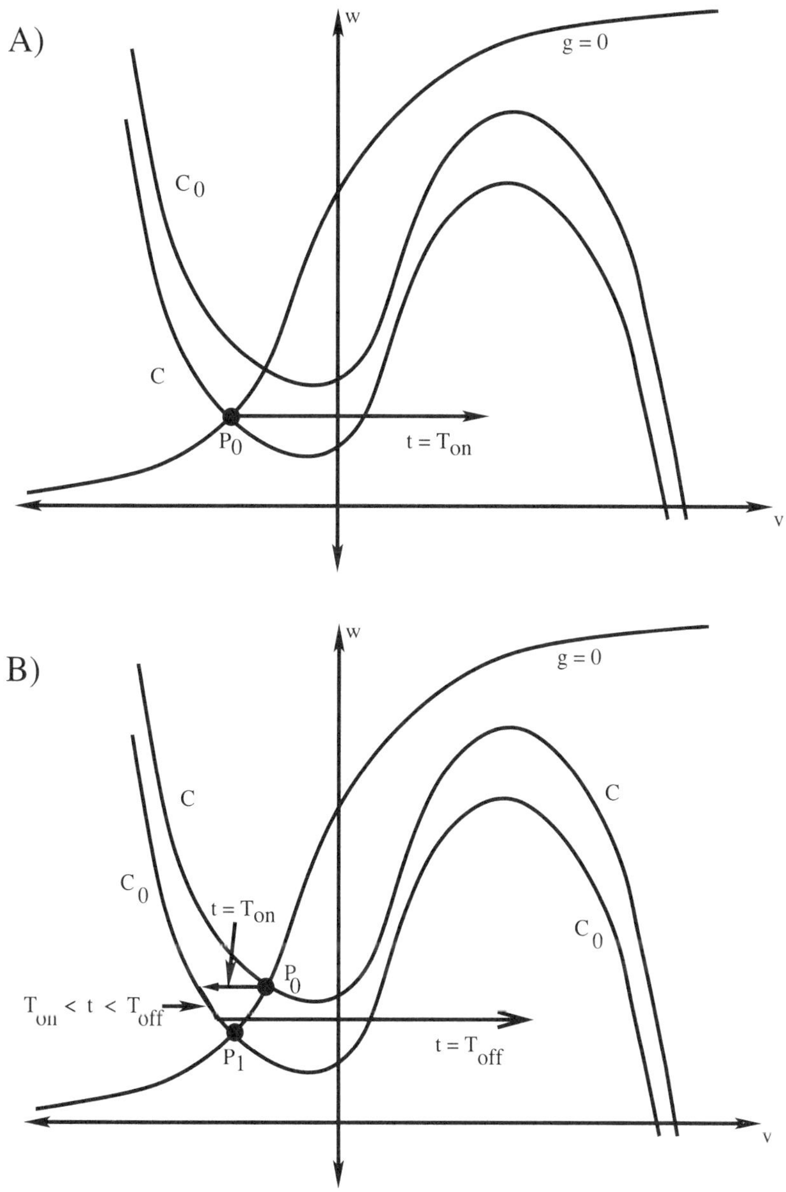

FIGURE 3. Phase space representation of the response of a model neuron to applied current. Current is applied at time $t = T_{on}$ and turned off at $t = T_{off}$. (A) Depolarizing current. The cell jumps up as soon as the current is turned on. (B) Hyperpolarizing current. The cell jumps to the left branch of C_0 when the current is turned on and jumps up to the active phase due to post-inhibitory rebound when the current is turned off.

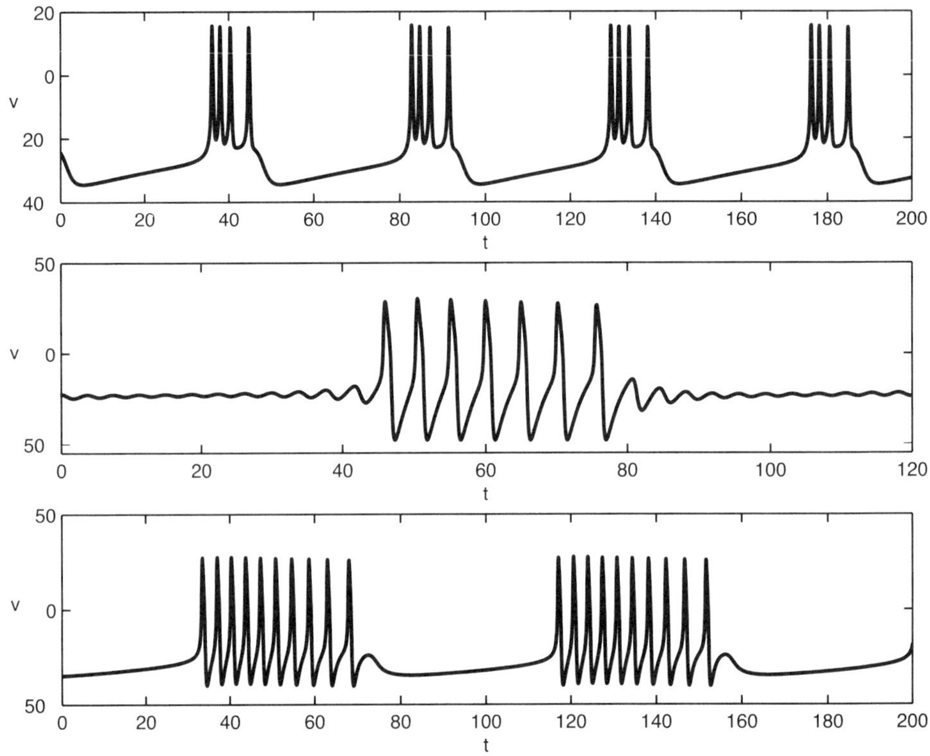

FIGURE 4. Classes of bursting oscillations. A. (top) Square-wave bursting. B. (middle) Elliptic bursting. C. (bottom) Parabolic bursting.

4. Bursting Oscillations

4.1. Introduction. Certain neurons and other excitable cells exhibit bursting oscillations; this behavior is characterized by a silent phase of near steady state resting behavior alternating with an active phase of rapid, spike-like oscillations, as shown in Figure 4. Examples of biological systems which display bursting oscillations include the Aplysia R-15 neuron, insulin secreting pancreatic beta cells, and neurons in the hippocampus, cortex and thalamus. For references, see, for example, [**59**].

Figure 4 shows three types of bursting oscillations. Figure 4A displays an example of *square-wave* bursting. This is characterized by abrupt periodic switching between the quiescent, or silent, phase and the active phase of repetitive firing. Note that the frequency of spikes decreases at the end of the active phase. Figure 4B illustrates *elliptic* bursting. Small amplitude oscillations occur during the silent phase and the amplitude of spikes gradually waxes and wanes. Finally, Figure 4C displays *parabolic* bursting. The spike rate first increases and then decreases in a parabolic manner.

The mathematical mechanisms responsible for each class of bursting oscillation are described in terms of geometric properties of the corresponding phase space dynamics. The model for each involves multiple time scales and can be written as

$$\begin{aligned} x' &= f(x,y) \\ y' &= \epsilon g(x,y) \end{aligned} \tag{4.1}$$

Here, $x \in R^n$ represents fast variables, while $y \in R^m$ represents slow variables; ϵ is a small singular perturbation parameter. We refer to the first equation in (4.1), with y considered as constant, as the fast subsystem (FS). The silent phase of the bursting solution corresponds to the passage of a trajectory of (4.1) near a manifold of fixed points of (FS). The active phase of repetitive spikes corresponds to the passage of the trajectory near a manifold of periodic solutions of (FS). It is the slow processes which modulate the fast dynamics between these two phases. Different classes of bursting oscillations are distinguished by the mechanisms by which the bursting trajectories switch between the silent and active phases. This is closely related to the global bifurcation structure of the fast subsystem with the slow variables treated as parameters.

We note that models for bursting oscillations may exhibit other types of periodic solutions, as well as more exotic behavior including chaotic dynamics [**7, 49**]. The models contain multiple time scales and this often leads to very interesting issues related to the theory of singular perturbations. For example, homoclinic orbits usually play an important role in the generation of these rhythms; the active phase of rapid oscillations may either begin or end (or both) as the bursting trajectory crosses a homoclinic point. At these points, standard singular perturbation methods may break down, so more delicate analysis is required. Moreover, the homoclinic orbits are often directly responsible for the generation of chaotic dynamics.

4.2. Square-Wave Bursting. Here we describe the mathematical mechanisms responsible for square-wave bursting. The analysis here follows Rinzel [**34**]. We consider a system of the form (4.1) and make geometric assumptions concerning the set of fixed points and periodic solutions of (FS). Some assumptions are then needed on the equations governing the slow variables. These assumptions can usually be verified for a specific system by using numerical software.

We only consider the simplest, lowest dimensional, models which generate square-wave bursting. This is a three-dimensional version of (4.1) of the form

$$\begin{aligned} v' &= f(v,w,y) \\ w' &= g(v,w,y) \\ y' &= \epsilon h(v,w,y,\lambda) \end{aligned} \tag{4.2}$$

The fast subsystem (FS) now consists of the first two equations in (4.2) with the slow variable y considered as constant. In (4.2), λ is a fixed parameter. We discuss how solutions of (4.2) may change with respect to λ in Remark 4.4.

The primary assumptions on (4.2) concern the bifurcation structure of the fast subsystem with y treated as a parameter. This structure appears in Figure 5, which also shows the projection of the square-wave bursting solution onto the corresponding bifurcation diagram. The set of fixed points of (FS) is assumed to be a Z-shaped curve in the (v, w, y) phase space. We denote this curve by S; only a portion of this Z-shaped curve is shown in Figure 5. The fixed points along the lower branch of S are stable solutions of (FS), while the fixed points on the middle branch of S are saddles. Fixed points along the upper branch of S may be stable or unstable. We also assume that there exists a one-parameter family of periodic solutions of (FS), denoted by P. These limit cycles originate at a (subcritical) Hopf bifurcation along the upper branch of S and terminate along a solution of (FS) that is homoclinic to one of the fixed points on the middle branch of S.

Assumptions are also needed about the slow dynamics. We assume that the y-nullsurface $\{h = 0\}$ defines a two-dimensional manifold that intersects S at a single point. This point lies on the middle branch of S between the homoclinic point and the left knee of S. Finally, $h > 0$ above $\{h = 0\}$ and $h < 0$ below $\{h = 0\}$.

We now give a heuristic explanation for why this system generates a square-wave bursting solution. Suppose that $\epsilon > 0$ is small and consider a solution that begins close to the lower branch. Because this branch consists of stable fixed points of (FS), the trajectory quickly approaches a small neighborhood of the lower branch. The trajectory tracks leftward along the lower branch according to the slow dynamics, until it passes the left knee. This portion of the solution corresponds to the silent phase. Once past the left knee, the trajectory is attracted to near P, the branch of periodic solutions of (FS). This generates the fast repetitive spikes of the bursting solutions. The trajectory passes near P, with increasing y, until it reaches a neighborhood of the homoclinic orbit of (FS). Once it passes the homoclinic orbit, the fast dynamics eventually forces the trajectory back to near the lower branch of S and this completes one cycle of the bursting solution.

This description is formal. It is not at all clear that if the system (4.2) satisfies the above assumptions, then, for all ϵ sufficiently small, there exists a unique periodic solution corresponding to a bursting oscillation. In fact, it is shown in [**49**] that such a result cannot be true, in general. However, in [**25**], it is proved that the bursting solution will be uniquely determined for all ϵ sufficiently small, except for those ϵ that lie in a certain very small set.

REMARK 4.1. A crucial ingredient for square-wave bursting is bistability. This allows for a hysteresis loop between a lower branch of stable fixed points and an upper branch of stable limit cycles. It is also very important that the slow nullsurface $\{h = 0\}$ lies between these two branches. If this last condition is not satisfied, then the system may exhibit other types of solutions. For example, suppose that $\{h = 0\}$ intersects the lower branch of S. This point of intersection will then be a globally stable fixed point of (4.2). If, on the other hand, $\{h = 0\}$ intersects S along its middle branch above the homoclinic point, then (4.2) may give rise to a stable limit cycle which always remains in the active phase near P. This type of solution is referred to as *continuous spiking*. Rigorous results concerning the existence of continuous spiking are presented in [**49, 50**]

REMARK 4.2. Square-wave bursting arises in models for electrical activity in pancreatic β-cells. It is believed that this activity plays an important role in the release of insulin from the cells. The first mathematical model for this bursting was due to Chay and Keizer [**6**]. There have been numerous related models, based on experimental data, since then. A review of these models, along with a detailed description of the more biological issues, is given in [**42**]. Square wave bursting also arises in recent models for respiratory CPG neurons [**4**] and models for pattern generation based on synaptic depression [**48**].

REMARK 4.3. It is still not clear what underlying biological mechanisms cause bursting oscillations in pancreatic β-cells [**8**]. In fact, it is possible that bursting may not be generated by the intrinsic properties of a single cell; rather, the bursting may be a network property generated by the coupling between cells. Computational

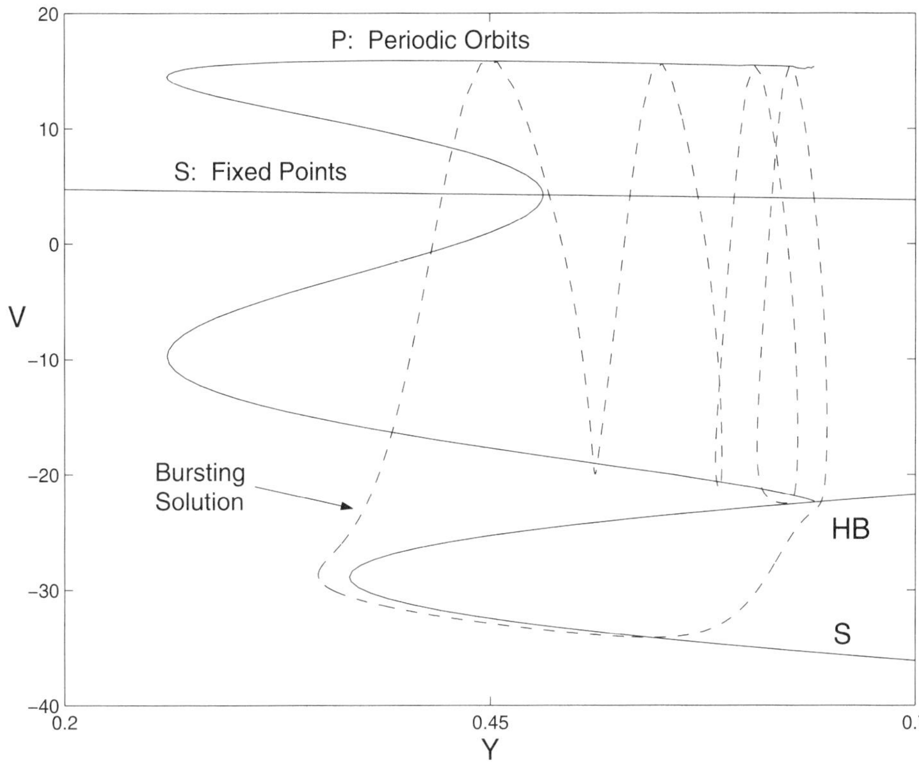

FIGURE 5. Bifurcation structure for square-wave bursting. The dashed trajectory is the projection of a square-wave bursting solution. Also shown are the branch of periodic solutions P, the branch of fixed points S and the Hopf bifurcation point HB.

studies in [**43**] have demonstrated that a population of cells with electrical, or gap-junction, coupling can exhibit bursting oscillations although the parameters of each cell are set so that the cells do not oscillate without any coupling. Detailed analysis of two mutually coupled square-wave bursters is given in [**41**].

REMARK 4.4. Very complicated (global) bifurcations can take place as the parameters ϵ or λ are varied in (4.2). The singular perturbation parameter ϵ controls the rate at which a bursting trajectory passes through the silent and active phases. In particular, the number of spikes per burst is $O(1/\epsilon)$ and becomes unbounded as $\epsilon \to 0$. It is demonstrated in [**49**] that Smale horseshoe chaotic dynamics can arise during the transition of adding a spike.

Perhaps even more interesting is the bifurcation structure of (4.2) as λ is varied. In the β-cell models, λ is related to the glucose concentration. As the glucose level gradually increases, the cells exhibit resting behavior, then bursting oscillations, and then continuous spiking. This is consistent with behavior exhibited by the model. As λ increases, the y-nullsurface $\{h = 0\}$ intersects the lower branch of S, then the middle branch of S below the homoclinic point, and then the middle branch of S above the homoclinic point. Numerical studies [**7**] and rigorous analysis [**50**] have shown that as λ varies between the bursting and continuous spiking regimes, the bifurcation structure of solutions must be very complicated. A Poincaré return map defined by the flow from a section transverse to the homoclinic orbit of (FS) will exhibit Smale-horseshoe dynamics for a robust range of parameter values. This leads to solutions in which the number of spikes per burst varies considerably.

REMARK 4.5. Some phenomenological, polynomial models for square-wave bursting have been proposed. See, for example, [**9**], [**14**], [**31**]. Analysis of models with two slow variables which exhibit square-wave bursting is given in [**44**].

REMARK 4.6. A system of equations which give rise to square-wave bursting is [**35**]:

$$\begin{aligned} v' &= -(g_{ca}m_\infty(v)(v-v_{ca}) + g_k w(v-v_k) + g_l(v-v_l) + g_{kca}z(y)(v-v_k)) + I \\ w' &= 20\phi(w_\infty(v)-w)/\tau(v) \\ y' &= 20\epsilon(-\mu g_{ca}m_\infty(v)(v-v_{ca}) - y) \end{aligned}$$

where, $g_{ca} = 4.$, $g_k = 8.0$, $g_l = 2.0$, $v_k = -84$, $v_l = -60$, $v_{ca} = 120.0$, $I = 45$, $g_{kca} = .25$, $\phi = .23$, $\epsilon = .005$, and $\mu = .02$. The nonlinear functions are given by $m_\infty(v) = .5(1. + \tanh((v+1.2)/18))$, $w_\infty(v) = .5(1. + \tanh((v-12)/17.4))$, $z(y) = y/(1+y)$ and $\tau(v) = \cosh((v-12.)/34.8)$.

4.3. Elliptic Bursting. Elliptic bursting can also arise in a system of the form (4.2) in which there are two fast variables and one slow variable. The bifurcation diagram of (FS) for an elliptic bursting scenario, with the slow variable treated as a parameter, is shown in Figure 6, which includes the projection of the elliptic bursting solution.

Bistability is crucially important for the generation of elliptic bursting, just as it is for square-wave bursting. An important difference, however, is that for elliptic bursting, the curve S of fixed points of (FS) need not be Z-shaped; there may be only one fixed point of (FS) for each value of y. The branch of periodic solutions P now originates at a subcritical Hopf bifurcation along S. To obtain bursting, we hypothesize that the slow variable decreases along S; that is, $h < 0$ near S. During the silent phase, the bursting solution evolves near the stable portion of S until it passes the Hopf point, beyond which the fixed points of S are no longer stable. Note, however, that the trajectory does not jump up to the active phase immediately after crossing the Hopf point. The slow variable y may traverse a distance that is $O(1)$ with respect to ϵ past the Hopf point before jumping up. This type of delayed behavior or slow-passage past a Hopf point has been studied extensively in the singular perturbation literature [**2, 30**].

We must also hypothesize that there is a net increase in the slow variable as the bursting trajectory passes near P. More precisely, let $(v_y(t), w_y(t))$ denote the periodic solutions along the outer branch of P and suppose that their periods are $T(y)$. Let

$$\bar{h}(y) = \frac{1}{T(y)}\int_0^{T(y)} h(v_y(t), w_y(t), y)dt$$

denote the average of h along these limit cycles. We assume that $\bar{h}(y) > 0$ for each y. Thus during the active phase, y increases until the bursting solution passes the turning point, or knee, of P. The fast dynamics then forces the trajectory back towards S and a new silent phase begins.

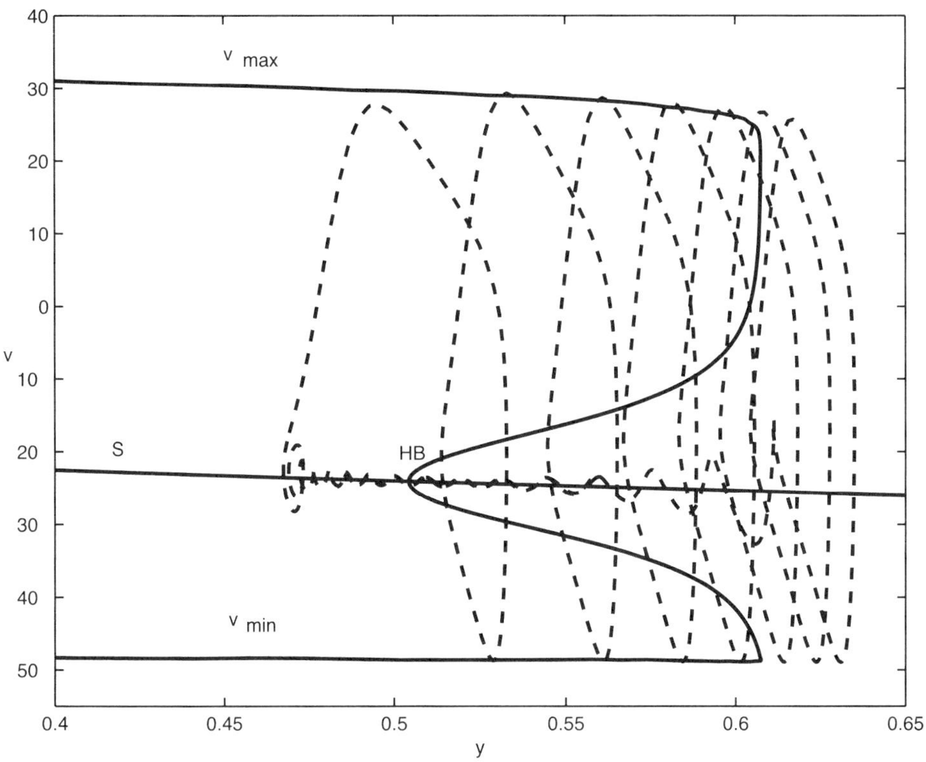

FIGURE 6. Bifurcation structure for elliptic bursting.

REMARK 4.7. Both square-wave and elliptic bursting depend on bistability and hysteresis. An important difference between these two classes of bursting is how the active phase terminates. Square-wave bursting ends at a homoclinic bifurcation; the period of oscillation, therefore, increases at the end of each burst. For elliptic bursting, the active phase ends at a saddle node of periodic orbits. There is no pattern for the spike frequencies in general.

REMARK 4.8. Elliptic bursting arises in models for thalamic neurons [**11**], rodent trigeminal neurons [**29**], and 40-Hz oscillations [**26**], [**57**]

REMARK 4.9. Here we assumed that the branch of periodic orbits of (FS) originate at a subcritical Hopf bifurcation. Hoppensteadt and Izhikevich [**16**] show that elliptic bursting is also possible even if the Hopf bifurcation is supercritical.

REMARK 4.10. A system of equations which give rise to elliptic bursting is [**35**]:

$$\begin{aligned} v' &= -(g_{ca}m_\infty(v)(v-v_{ca}) + g_k w(v-v_k) + g_l(v-v_l) + g_{kca}z(y)(v-v_k)) + I \\ w' &= \phi(w_\infty(v) - w)/\tau(v) \\ y' &= \epsilon(-\mu g_{ca}m_\infty(v)(v-v_{ca}) - y) \end{aligned}$$

where, $g_{ca} = 4.4$, $g_k = 8.0$, $g_l = 2.0$, $v_k = -84$, $v_l = -60$, $v_{ca} = 120.0$, $I = 120$, $g_{kca} = .75$, $\phi = 1.2$, $\epsilon = .04$, and $\mu = .016667$. The nonlinear functions are given by $m_\infty(v) = .5(1. + \tanh((v + 1.2)/18))$, $w_\infty(v) = .5(1. + \tanh((v - 2.)/30.))$, $z(y) = y/(1 + y)$ and $\tau(v) = \cosh((v - 2.)/60.)$.

4.4. Parabolic Bursting. Both square-wave and elliptic bursting can be achieved in a system with only one slow variable. Moreover, both depend on bistability of the fast dynamics. Parabolic bursting, on the other hand, requires at least two slow variables and does not arise from a hysteresis phenomenon. A geometric model for parabolic bursting is the following. Consider a system of the form (4.1) where $x \in R^2$ and $y = (y_1, y_2) \in R^2$. There are now two slow variables, namely y_1 and y_2. We first describe the bifurcation structure of the fast subsystem with both slow variables considered as parameters.

We hypothesize that the set of fixed points of (FS) forms a Z-shaped surface. The fixed points along the lower branch of this surface are assumed to be stable fixed points of (FS). There is a curve along the upper branch of fixed points where Hopf bifurcations occur. Periodic solutions arise at these Hopf points and terminate along orbits homoclinic to the fixed points along the lower fold of the fixed point surface.

The existence of a parabolic bursting solution also requires hypotheses on the slow dynamics. To write reduced equations which determine the evolution of the slow variables in the silent phase, denote the lower branch of fixed points by $x = \Phi(y)$. The silent phase evolution of the slow variables is then given by the reduced equations

$$\dot{y} = g(\Phi(y), y) \tag{4.3}$$

where differentiation is now with respect to $\tau = \epsilon t$.

One obtains reduced equations for the evolution of the slow variables in the active phase using averaging. Suppose that $y = (y_1, y_2)$ lies in the region where there exists a stable limit cycle of (FS). Let $x_y(t)$ be the corresponding periodic solution of (FS) with period $T(y)$ and consider the averaged quantity

$$\bar{g}(y) = \frac{1}{T(y)} \int_0^{T(y)} g(x_y(t), y)dt$$

The active phase evolution of the slow variables is then given by the averaged equations

$$\dot{y} = \bar{g}(y) \tag{4.4}$$

Parabolic bursting corresponds to the existence of a closed curve in the slow (y_1, y_2) phase plane which passes through both the region of stable fixed points of (FS), where it satisfies (4.3), and the region of stable limit cycles of (FS), where it satisfies (4.4).

Note that the active phase of the bursting solution both begins and ends along a curve of homoclinic bifurcations. Since the periods of the limit cycles tend to infinity at the homoclinic bifurcations, the interspike interval is longer at both the beginning and end of each burst. This accounts for the parabolic nature of the period of fast oscillations.

REMARK 4.11. Parabolic bursting is found in the Aplysia R-15 neuron [**1**]. Rinzel and Lee [**36**] considered a model due to Plant [**33**] and were the first to give a detailed analysis of parabolic bursting by dissecting the equations into fast and slow subsystems as described here.

REMARK 4.12. Rigorous results related to parabolic bursting are given in [**23**], [**47**], and [**16**].

REMARK 4.13. A system of equations which give rise to parabolic bursting is [**35**]:

$$\begin{aligned} v' &= -(i_{ca}(v) + i_k(v) + i_l(v) + i_{kca}(v,c) + i_{cas}(v,s)) + I \\ w' &= \phi(w_\infty(v) - w)/\tau_w(v) \\ c' &= \epsilon(-\mu g_{ca} m_\infty(v)(v - v_{ca}) - c) \\ s' &= \epsilon(s_\infty(v) - s)/\tau_s \end{aligned}$$

where, $i_{ca}(v) = g_{ca} m_\infty(v)(v - v_{ca})$, $i_k(v) = g_k w(v - v_k)$, $i_l(v) = g_l(v - v_l)$, $i_{kca}(v,c) = g_{kca} z(c)(v - v_k)$, and $i_{cas}(v,s) = g_{cas} s(v - v_{ca})$. The constants are given by $g_{ca} = 4.$, $g_k = 8.0$, $g_l = 2.0$, $v_k = -84$, $v_l = -60$, $v_{ca} = 120.0$, $I = 65$, $g_{kca} = 1.$, $\phi = 1.333$, $\epsilon = .002$, $\mu = .025$, $\tau_s = .05$, and $g_{cas} = 1.$. The nonlinear functions are given by $m_\infty(v) = .5(1. + \tanh((v + 1.2)/18))$, $w_\infty(v) = .5(1. + \tanh((v - 12.)/17.))$, $\tau_w(v) = \cosh((v - 12.)/34.)$, $z(c) = c/(1 + c)$, and $s_\infty = .5(1 + \tanh((v - 12/24))$.

5. Two Mutually Coupled Cells

5.1. Introduction. In this section, we consider a network consisting simply of two mutually coupled cells. By considering such a simple system, we are able to describe how we model networks of oscillators, the types of behavior that can arise in such systems and the mathematical techniques we use for the analysis of the behavior. For this discussion, we assume that each cell, without any coupling, is modeled as the relaxation oscillator

$$\begin{aligned} v' &= f(v,w) \\ w' &= \epsilon g(v,w) \end{aligned} \tag{5.1}$$

Here ϵ is assumed to be small; that is, w represents a slowly evolving quantity. As in Section 3, we assume that the v-nullcline, $f(v,w) = 0$, defines a cubic-shaped curve and the w-nullcline, $g = 0$, is a monotone decreasing curve which intersects $f = 0$ at a unique point p_0. We also assume that $f > 0$ ($f < 0$) below (above) the v-nullcline and $g > 0$ (< 0) below (above) the w-nullcline.

System (5.1) can be viewed as a simple model for a bursting neuron in which the active phase corresponds to the envelope of a burst's rapid spikes. Of course, a two-dimensional model for a single cell cannot exhibit the more exotic dynamics described in the previous section for a bursting cell. However, by considering a simple relaxation-type oscillator for each cell, we will be able to discuss how network properties contribute to the emergent behavior of a population of cells. It is, of course, a very interesting issue to understand how this population behavior changes when one considers more detailed models for each cell. Some results for more detailed models are given in [**40**].

In the next section, we describe how we model the two mutually coupled cells. The form of coupling used is referred to as *synaptic coupling* and is meant to correspond to a simple model for chemical synapses. As described earlier, there are many different forms of synaptic coupling. For example, it may be excitatory or inhibitory and it may exhibit either fast or slow dynamics. We are particularly interested in how the nature of the synaptic coupling affects the emergent population rhythm. A natural question is whether excitatory or inhibitory coupling leads to either synchronous or asynchronous rhythms. There are four possible combinations and all four may, in fact, be stably realized, depending on the details of the intrinsic and synaptic properties of the cells. Here we discuss conditions for when excitatory coupling leads to synchronous rhythms and inhibitory coupling leads to antiphase behavior. Many of the results given here are complementary to those in [**22**] who use the slow structure in a somewhat different way.

5.2. Synaptic Coupling. We model a pair of mutually coupled neurons by the following system of differential equations

$$\begin{aligned} v_1' &= f(v_1, w_1) - s_2 g_{syn}(v_1 - v_{syn}) \\ w_1' &= \epsilon g(v_1, w_1) \\ v_2' &= f(v_2, w_2) - s_1 g_{syn}(v_2 - v_{syn}) \\ w_2' &= \epsilon g(v_2, w_2) \end{aligned} \tag{5.2}$$

Here (v_1, w_1) and (v_2, w_2) correspond to the two cells. The coupling term $s_j g_{syn}(v_i - v_{syn})$ can be viewed as an additional current which may change a cell's membrane potential v_i. The parameter g_{syn} corresponds to the maximal conductance of the synapse and is positive, while the reversal potential v_{syn} determines whether the synapse is excitatory or inhibitory. If $v < v_{syn}$ along each bounded singular solution, then the synapse is excitatory, while if $v > v_{syn}$ along each bounded singular solution, then the synapse is inhibitory.

The terms s_i, $i = 1, 2$, in (5.2) encode how the postsynaptic conductance depends on the presynaptic potentials v_i. There are several possible choices for the s_i. The simplest choice is to assume that $s_i = H(v_i - \theta_{syn})$, where H is the Heaviside step function and θ_{syn} is a threshold above which one cell can influence the other. Note, for example, that if $v_1 < \theta_{syn}$, then $s_1 = H(v_1 - \theta_{syn}) = 0$, so cell 1 has no influence on cell 2. If, on the other hand, $v_1 > \theta_{syn}$, then $s_1 = 1$ and cell 2 is affected by cell 1.

Another choice for the s_i is to assume that they satisfy a first order equation of the form

$$s_i' = \alpha(1 - s_i)H(v_i - \theta_{syn}) - \beta s_i \tag{5.3}$$

where α and β are positive constants and H and θ_{syn} are as before. Note that α and β are related to the rates at which the synapses turn on or turn off. For *fast synapses*, we assume that both of these constants are $O(1)$ with respect to ϵ. For a *slow synapse*, we assume that $\alpha = O(1)$ and $\beta = O(\epsilon)$; hence, a slow synapse activates on the fast time scale but turns off on the slow time scale.

The synapses considered so far are referred to as *direct* synapses since they are activated as soon as a membrane potential crosses the threshold θ_{syn}. To more fully represent the range of synapse dynamics observed biologically, it is also necessary to consider more complicated connections. These are referred to as *indirect synapses*, and they are modeled by introducing new dependent variables x_1 and x_2. Each (x_i, s_i) satisfies the equations

$$
\begin{aligned}
x_i' &= \epsilon\alpha_x(1 - x_i)H(v_i - \theta_v) - \epsilon\beta_x x_i \\
s_i' &= \alpha(1 - s_i)H(x_i - \theta_x) - \beta s_i
\end{aligned}
\tag{5.4}
$$

The constants α_x and β_x are assumed to be independent of ϵ. The effect of the indirect synapses is to introduce a delay from the time one oscillator jumps up until the time the other oscillator feels the synaptic input. For example, if the first oscillator jumps up, a secondary process is turned on when v_1 crosses the threshold θ_v. The synapse s_1 does not turn on until x_1 crosses θ_x; this takes a finite amount of (slow) time since x_1 evolves on the slow time scale, like the w_i. Note that indirect synapses may be fast or slow. For fast, indirect synapses, the turn off of s_i, after the x_i-induced delay, occurs on the fast time scale.

5.3. Geometric Approach. All of the networks in this paper are analyzed by treating ϵ as a small, singular perturbation parameter. As in the previous section, the first step in the analysis is to identify the fast and slow variables. We then dissect the full system of equations into fast and slow subsystems. The fast subsystem is obtained by simply setting $\epsilon = 0$ in the original equations. This leads to a reduced set of equations for the fast variables with each of the slow variables held constant. The slow subsystems are obtained by first introducing the slow time scale $\tau = \epsilon t$ and then setting $\epsilon = 0$ in the resulting equations. This leads to a reduced system of equations for just the slow variables, after solving for each fast variable in terms of the slow ones. The slow subsystems determine the evolution of the slow variables while the cells are in either the active or the silent phase. During this time, each cell lies on either the left or the right branch of some "cubic" nullcline determined by the total synaptic input which the cell receives. This continues until one of the cells reaches the left or right "knee" of its corresponding cubic. Upon reaching a knee, the cell may either jump up from the silent to the active phase or jump down from the active to the silent phase. The jumping up or down process is governed by the fast equations.

For a concrete example, consider two mutually coupled cells with fast, direct synapses. The dependent variables (v_i, w_i, s_i), $i = 1, 2$, then satisfy (5.2) and (5.3). The slow equations are

$$
\begin{aligned}
0 &= f(v_i, w_i) - s_j g_{syn}(v_i - v_{syn}) \\
\dot{w}_i &= g(v_i, w_i) \\
0 &= \alpha(1 - s_i)H(v_i - \theta_{syn}) - \beta s_i
\end{aligned}
\tag{5.5}
$$

where differentiation is with respect to τ and $i \neq j$. The first equation in (5.5) states that (v_i, w_i) lies on a curve determined by s_j. The third equation states that if cell i is silent ($v_i < \theta_{syn}$), then $s_i = 0$, while if cell i is active, then $s_i = \frac{\alpha}{\alpha+\beta} \equiv s_A$. We demonstrate that it is possible to reduce (5.5) to a single equation for each of the slow variables w_i. Before doing this, it will be convenient to introduce some notation.

Let $\Phi(v,w,s) \equiv f(v,w) - g_{syn}s(v - v_{syn})$. If g_{syn} is not too large, then each $C_s \equiv \{\Phi(v,w,s) = 0\}$ defines a cubic-shaped curve. We express the left and right branches of C_s by $\{v = \Phi_L(w,s)\}$ and $\{v = \Phi_R(w,s)\}$, respectively. Finally, let

$$G_L(w,s) \;=\; g(\Phi_L(w,s),w) \text{ and } G_R(w,s) \;=\; g(\Phi_R(w,s),w)$$

Now the first equation in (5.5) can be written as $0 = \Phi(v_i, w_i, s_j)$ with s_j fixed. Hence, $v_i = \Phi_\alpha(v_i, s_j)$ where $\alpha = L$ if cell i is silent and $\alpha = R$ if cell i is active. It then follows that each slow variable w_i satisfies the single equation

$$\dot{w}_i \;=\; G_\alpha(w_i, s_j) \tag{5.6}$$

By dissecting the full system into fast and slow subsystems, we are able to construct singular solutions of (5.2),(5.3). In particular, this leads to sufficient conditions for when there exists a singular synchronous solution and when this solution is (formally) asymptotically stable. The second step in the analysis is to rigorously prove that the formal analysis, in which $\epsilon = 0$, is justified for small $\epsilon > 0$. This raises some very subtle issues in the geometric theory of singular perturbations, some of which have not been completely addressed in the literature. For most of the results presented here, we only consider singular solutions.

We note that the geometric approach used here is somewhat different from that used in many dynamical systems studies (see, for example, [**35**]). All of the networks considered here consist of many differential equations, especially for larger networks. Traditionally, one would interpret the solution of this system as a single trajectory evolving in a very large dimensional phase space. We consider several trajectories, one corresponding to a single cell, moving around in a much lower dimensional phase space (see also [**40**], [**45**], [**52**], [**53**], [**55**]). After reducing the full system to a system for just the slow variables, the dimension of the lower dimensional phase space equals the number of slow intrinsic variables and slow synaptic variables corresponding to each cell. In the worst case considered here, there is only one slow intrinsic variable for each cell and one slow synaptic variable; hence, we never have to consider phase spaces with dimension more than two. Of course, the particular phase space we need to consider may change, depending on whether the cells are active or silent and also depending on the synaptic input that a cell receives.

5.4. Synchrony With Excitatory Synapses. Consider two mutually coupled cells with excitatory synapses. Our goal here is to give sufficient conditions for the existence of a synchronous solution and its stability. Note that if the synapses are excitatory, then the curve $C_A \equiv C_{s_A}$ lies 'above' $C_0 \equiv \{f = 0\}$ as shown in Figure 7. This is because for an excitatory synapse, $v < v_{syn}$ along the synchronous solution. Hence, on C_A, $f(v,w) = g_{syn}s_A(v - v_{syn}) < 0$, and we are assuming that $f < 0$ above C_0. If g_{syn} is not too large, then both C_0 and C_A will be cubic shaped. We assume that the threshold θ_{syn} lies between the two knees of C_0. In the statement of the following result, we denote the left knee of C_0 by (v_{LK}, w_{LK}).

THEOREM 5.1. *Assume that each cell, without any coupling, is oscillatory. Moreover, assume the synapses are fast, direct and excitatory. Then there exists a synchronous periodic solution of (5.2), (5.3). This solution is asymptotically stable if one of the following two conditions is satisfied.*

(H1) $\frac{\partial f}{\partial w} < 0$, $\frac{\partial g}{\partial v} > 0$, and $\frac{\partial g}{\partial w} < 0$ *near the singular synchronous solution.*

(H2) $|g(v_{LK}, w_{LK})|$ *is sufficiently small.*

REMARK 5.2.. We note that the synchronous solution cannot exist if the cells are excitable and the other hypotheses, concerning the synapses, are satisfied. This is because along a synchronous solution, each (v_i, w_i) lies on the left branch of C_0 during the silent phase. If the cells are excitable, then each (v_i, w_i) will approach the point where the w-nullcline $\{g = 0\}$ intersects the left branch of C_0. The cells, therefore, will not be able to jump up to the active phase.

REMARK 5.3.. The assumptions concerning the partial derivatives of f and g in (H1) are not very restrictive since we are already assuming that $f > 0$ (< 0) below (above) the v-nullcline and $g > 0$ (< 0) below (above) the w-nullcline.

REMARK 5.4.. A useful way to interpret (H2) is that the silent phases of the cells are much longer than their active phases. This is because $g(v_{LK}, w_{LK})$ gives the rate at which the slow variables w_i evolve near the end of the silent phase. Note that $g(v_{LK}, w_{LK})$ will be small if the left knee of C_0 is very close to the w-nullcline.

Proof: We first consider the existence of the synchronous solution. This is straightforward because along a synchronous solution $(v_1, w_1, s_1) = (v_2, w_2, s_2) \equiv (v, w, s)$ satisfy the reduced system

$$\begin{aligned} v' &= f(v, w) - s g_{syn}(v - v_{syn}) \\ w' &= \epsilon g(v, w) \\ s' &= \alpha(1 - s)H(v - \theta_{syn}) - \beta s \end{aligned}$$

The singular solution consists of four pieces. During the silent phase, $s = 0$ and (v, w) lies on the left branch of C_0. During the active phase $s = s_A$ and (v, w) lies on the right branch of C_A. The jumps between these two phases occur at the left and right knees of the corresponding cubics.

We next consider the stability of the synchronous solution to small perturbations. We begin with both cells close to each other in the silent phase on the left branch of C_0, with cell 1 at the left knee ready to jump up. We follow the cells around in phase space by constructing the singular solution until one of the cells returns to the left knee of C_0. As before, the singular solution consists of four pieces. We need to show that the cells are closer to each other after this complete cycle than they were initially.

The first piece of the singular solution begins when cell 1 jumps up. When $v_1(t)$ crosses θ_{syn}, $s_1(t) \to s_A$. This raises the cubic corresponding to cell 2 from C_0 to C_A. If $|w_1(0) - w_2(0)|$ is sufficiently small, corresponding to a sufficiently small perturbation, then cell 2 lies below the left knee of C_A. The fast equations then force cell 2 to also jump up to the active phase, as shown in Figure 7. Note that this piece takes place on the fast time scale. Hence, on the slow time scale, both cells jump up at precisely the same time.

During the second piece of the singular solution, both oscillators lie in the active phase on the right branch of C_A. Note that the ordering in which the oscillators track along the left and right branches has been reversed. While in the silent phase, cell 1 was ahead of cell 2. In the active phase, cell 2 leads the way. The oscillators remain on the right branch of C_A until cell 2 reaches the right knee.

The oscillators then jump down to the silent phase. Cell 2 is the first to jump down. When $v_2(t)$ crosses θ_{syn}, s_2 switches from s_A to 0 on the fast time scale.

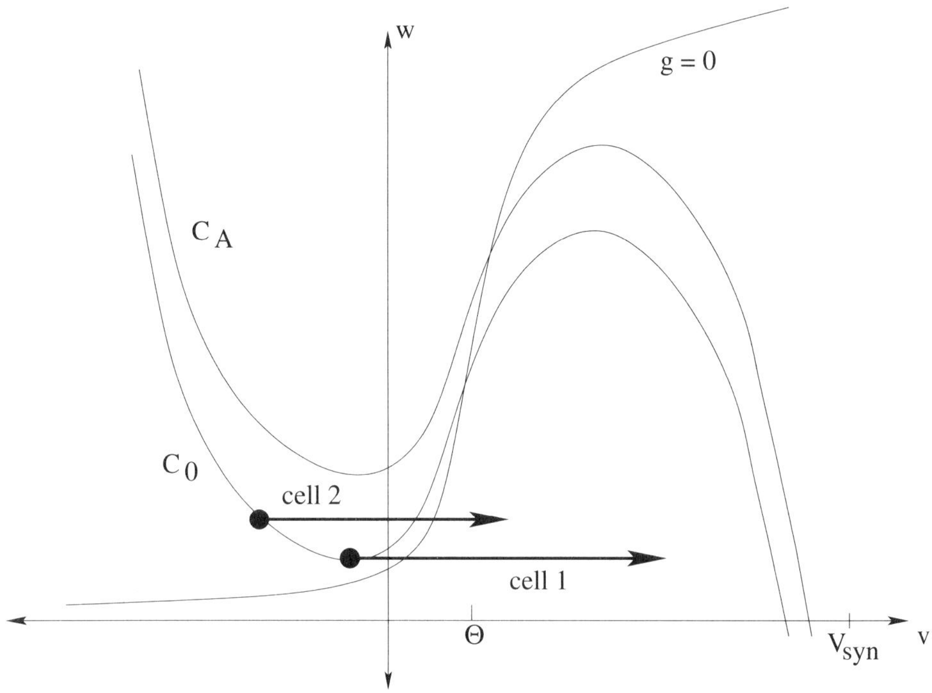

FIGURE 7. Nullclines for an oscillatory relaxation oscillator with (C_A) and without (C_0) excitatory coupling. Note that cell 2 responds to cell 1 through Fast Threshold Modulation.

This lowers the cubic corresponding to cell 1 from C_A to C_0. If, at this time, cell 1 lies above the right knee of C_A, then cell 1 must jump down to the silent phase. This will certainly be the case if the cells are initially close enough to each other.

During the final piece of the singular solution, both oscillators move down the left branch of C_0 until cell 1 reaches the left knee. This completes one full cycle.

To prove that the synchronous solution is stable, we must show that the cells are closer to each other after this cycle; that is, there is compression in the distance between the cells. There are actually several ways to demonstrate this compression; these correspond to two different ways to define what is meant by the 'distance' between the cells. Here we consider a Euclidean metric, which is defined as follows: Suppose that both cells lie on the same branch of the same cubic and the coordinates of cell i are (v_i, w_i). Then the distance between the cells is defined as simply $|w_1 - w_2|$. Note that during the jump up and the jump down, this metric remains invariant. This is because the jumps are horizontal so the values of w_i do not change. If there is compression, therefore, it must take place as the cells evolve in the silent and active phases. We now show that this is indeed the case if (H1) is satisfied.

Suppose that when $\tau = 0$, both cells lie in the silent phase on C_0. We assume, for convenience, that $w_2(0) > w_1(0)$. We need to prove that $w_2(\tau) - w_1(\tau)$ decreases as long as the cells remain in the silent phase. Now each w_i satisfies (5.6) with $\alpha = L$

and $s_j = 0$. Hence,

$$w_i(\tau) \;=\; w_i(0) \;+\; \int_0^\tau G_L(w_i(\xi), 0)d\xi$$

and, using the Mean Value Theorem,

(5.7)
$$\begin{aligned} w_2(\tau) - w_1(\tau) \;&=\; w_2(0) - w_1(0) \;+\; \textstyle\int_0^\tau \; G_L(w_2(\xi), 0) - G_L(w_1(\xi), 0)\; d\xi \\ &=\; w_2(0) - w_1(0) \;+\; \textstyle\int_0^\tau \; \frac{\partial G_L}{\partial w}(w^*, 0)(w_2(\xi) - w_1(\xi))\; d\xi \end{aligned}$$

for some w^*. Now $G_L(w,s) = g(\Phi_L(w), w)$. Hence, $\frac{\partial G_L}{\partial w} \;=\; g_v \Phi_L'(w) + g_w$. We assume in (H1) that $g_v > 0$ and $g_w < 0$ near the synchronous solution. Moreover, $\Phi_L'(w) < 0$ because $v = \Phi_L(w)$ defines the left branch of the cubic C_0 which has negative slope. It follows that $\frac{\partial G_L}{\partial w} < 0$, and therefore, from (5.7), $w_2(\tau) - w_1(\tau) < w_2(0) - w_1(0)$. This gives the desired compression; a similar computation applies in the active phase. We note that if there exists $\gamma > 0$ such that $\frac{\partial G_L}{\partial w} < -\gamma$ along the left branch, then Gronwall's inequality shows that $w_2(\tau) - w_1(\tau)$ decreases at an exponential rate.

We next consider (H2) and demonstrate why this leads to compression of trajectories. Suppose, for the moment, that $g(v_{LK}, w_{LK}) = 0$; that is, the left knee of C_0 touches the w-nullcline at some fixed point. Then both cells will approach this fixed point as they evolve along the left branch of C_0 in the silent phase. There will then be an infinite amount of compression, since both cells approach the same fixed point. It follows that we can assume that the compression is as large as we please by making $g(v_{LK}, w_{LK})$ sufficiently small. If the compression is sufficiently large, then it will easily dominate any possible expansion over the remainder of the cells' trajectories. This will, in turn, lead to stability of the synchronous solution.

REMARK 5.5. The mechanism by which one cell fires, and thereby raises the cubic of the other cell such that it also fires, was referred to as *Fast Threshold Modulation (FTM)* in [**45**]. There, a time metric was introduced to establish the compression of trajectories of excitatorily coupled cells, which implies the stability of the synchronou s solution. A detailed discussion of the time metric can be found in [**22**]; see also [**27**].

REMARK 5.6. While the synchronous solution has been shown to be stable, it need not be globally stable. In [**24**], it is shown that this network may exhibit stable antiphase solutions if certain assumptions on the parameters and nonlinear functions are satisfied.

We have so far considered a completely homogeneous network with just two cells. The analysis generalizes to larger inhomogeneous networks in a straightforward manner, if the degree of heterogeneity between the cells is not too large. The major difference in the analysis is that, with heterogeneity, the cells may lie on different branches of different cubics during the silent and active phases. The resulting solution cannot be perfectly synchronous; however, as demonstrated in [**55**], one can often expect synchrony in the jump-up, but not in the jump-down. Related work on heterogeneous networks include [**5**], [**32**], [**46**].

One may also consider, for example, an arbitrarily large network of identical oscillators with nearest neighbor coupling. We do not assume that the strength of coupling is homogeneous. Suppose that we begin the network with each cell in the silent phase. If the cells are identical, then they must all lie on the left branch of C_0. Now if one cell jumps up it will excite its neighbors and raise their corresponding cubics. If the cells begin sufficiently close to each other, then these neighbors will jump up due to FTM. In a similar manner, the neighbor's neighbors will also jump due to FTM and so on until every cell jumps up. In this way, every cell jumps up at the same (slow) time. While in the active phase, the cells may receive different input and, therefore, lie on the right branches of different cubics. Once one of the cells jumps down, there is no guarantee that other cells will also jump down at this (slow) time, because the cells to which it is coupled may still receive input from other active cells. Hence, one cannot expect synchrony in the jumping down process. Eventually every cell must jump down. Note that there may be considerable expansion in the distance between the cells in the jumping down process. If $|g(v_{LK}, w_{LK})|$ is sufficiently small, however, as in the previous result, then there will be enough compression in the silent phase so that the cells will still jump up together. Here we assumed that the cells are identical; however, the analysis easily follows if the heterogeneities among the cells are not too large. A detailed analysis of this network is given in [**55**].

5.5. Desynchrony With Inhibitory Synapses. We now consider two mutually coupled cells with inhibitory synapses. Under this coupling, the curve C_A now lies below C_0. As before, we assume that g_{syn} is not too large, such that both C_0 and C_A are cubic shaped. We also assume that the right knee of C_A lies above the left knee of C_0 as shown in Figure 8. Some assumptions on the threshold θ_{syn} are also required. For now, we assume that θ_{syn} lies between the left knee of C_0 and right knee of C_A.

We will assume throughout this section that the synapses are fast, direct and inhibitory. The main results state that if a synchronous solution exists then it must be unstable. The network will typically exhibit either out-of-phase oscillations or a completely quiescent state and we give sufficient conditions for when either of these arises. We note that the network may exhibit bistability; both the out-of-phase and completely quiescent solutions may exist and be stable for the same parameter values. These results are all for singular solutions. Some rigorous results for $\epsilon > 0$ are given in [**53**].

The first result concerns the existence and stability of the synchronous solution.

THEOREM 5.7. *Assume that the synapses are fast, direct and inhibitory. If each cell, without any coupling, is oscillatory and θ_{syn} is sufficiently large, then there exists a singular synchronous solution. This solution is unstable. If each cell, without any coupling, is excitable, then there does not exist a singular synchronous solution.*

Proof: The existence of a singular synchronous solution for oscillatory cells follows precisely as in the previous section. During the silent phase, the trajectory lies on the left branch of C_0, while in the active phase it lies on the right branch of C_A. Note that we require that the right knee of C_A lies above the left knee of C_0. Moreover, when the synchronous solution jumps up and crosses the threshold $v = \theta_{syn}$, it should lie to the right of the middle branch of C_A; otherwise, it would fall down to the silent phase. This is why we assume that θ_{syn} is sufficiently large.

This solution is unstable for the following reason. Suppose both cells are initially very close to each other on C_0. The cells then evolve on C_0 until one of the cells, say cell 1, reaches the left knee of C_0. Cell 1 then jumps up to the active phase. When v_1 crosses the threshold θ_{syn}, s_1 switches from 0 to s_A and cell 2 jumps from C_0 to C_A, as shown in Figure 8. This demonstrates that the cells are uniformly separated for arbitrarily close initial data. The synchronous solution must, therefore, be unstable.

The synchronous solution cannot exist if the cells are excitable for precisely the same reason discussed in the previous section. If such a solution did exist then each cell would lie on C_0 during its silent phase. Each cell would then approach the stable fixed point on this branch and would never be able to jump up to the active phase.

We next consider out-of-phase oscillatory behavior. One interesting feature of mutually coupled networks is that such oscillations can arise even if each cell is excitable for fixed levels of synaptic input. The following theorem gives sufficient conditions for when this occurs. We will require that the active phase of the oscillation is sufficiently long. To give precise conditions, we introduce the following notation.

Assume that the left and right knees of C_0 are at (v_{LK}, w_{LK}) and (v_{RK}, w_{RK}), respectively. If the w-nullcline intersects the left branch of C_A, then we denote this point by $(v_A, w_A) = p_A$. We assume that $w_A < w_{LK}$. Let τ_L be the (slow) time it takes for the solution of (5.6) with $\alpha = L$ and $s = s_A$ to go from $w = w_{RK}$ to $w = w_{LK}$, and let τ_R be the time it takes for the solution of (5.6) with $\alpha = R$ and $s = 0$ to go from $w = w_{LK}$ to $w = w_{RK}$. Note that τ_L is related to the time a solution spends in the silent phase, while τ_R is related to the time a solution spends in the active phase.

THEOREM 5.8. *Assume that the cells are excitable for each fixed level of synaptic input and the synapses are fast, direct, and inhibitory. Moreover, assume that* $w_A < w_{LK}$ *and* $\tau_L < \tau_R$. *Then the network exhibits stable out-of-phase oscillatory behavior.*

REMARK 5.9. We do not claim that the out-of-phase solution is uniquely determined or that it corresponds to antiphase behavior. These results may hold; however, their proofs require more analysis than that given here.

REMARK 5.10. The rest state with each cell at the fixed point on C_0 also exists and is stable. Hence, if the hypotheses of Theorem 5.8 are satisfied, then the network exhibits bistability.

Proof: Suppose that we begin with cell 1 at the right knee of C_0 and cell 2 on the left branch of C_A with $w_A < w_2(0) < w_{LK}$. Then cell 1 jumps down and, when v_1 crosses the threshold θ_{syn}, cell 2's cubic switches from C_A to C_0. Since $w_2(0) < w_{LK}$, cell 2 lies below the left knee of C_0, so it must jump up to the active phase. After these jumps, cell 1 lies on the left branch of C_A, while cell 2 lies on the right branch of C_0.

Cell 2 then moves up the right branch of C_0 while cell 1 moves down the left branch of C_A, approaching p_A. This continues until cell 2 reaches the right knee of C_0 and jumps down. We claim that at this time, cell 1 lies below the left knee of C_0, so it must jump up. We can then keep repeating this argument to obtain the

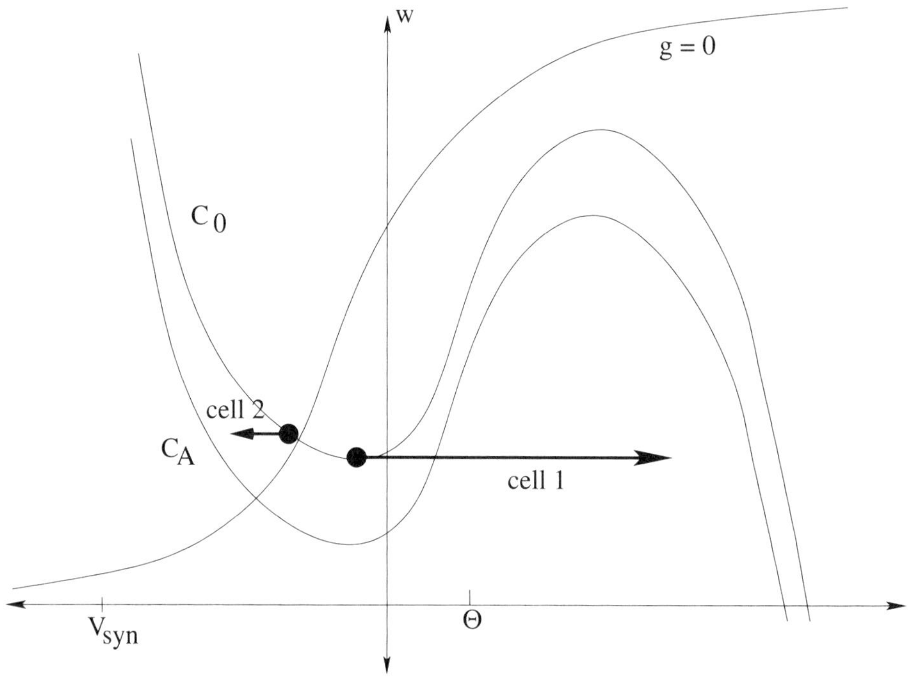

FIGURE 8. Instability induced by mutual inhibition. Cell 2 jumps to C_A when cell 1 fires.

sustained out-of-phase oscillations. The reason why cell 1 lies below the left knee of C_0 when cell 2 jumps down is because it spends a sufficiently long amount of time in the silent phase. To estimate this time, note that because cell 2 was initially below the left knee of C_0, the time it spends in the active phase before jumping down is greater than τ_R. Hence, the time cell 1 spends in the silent phase from the time it jumps down is greater than $\tau_R > \tau_L$. From the definitions, since cell 1 was initially at the right knee of C_0, it follows that cell 1 must be below the left knee of C_0 when cell 2 jumps down, which is what we wished to show.

REMARK 5.11. To obtain sustained oscillations, it is not really necessary to assume that each cell is excitable for all levels of synaptic input. Suppose, for example, that the w-nullcline intersects C_0 along its middle branch, but it intersects C_A along its left branch. Then each cell, without any coupling, is oscillatory. The hypothesis $\tau_L < \tau_R$ is no longer necessary for sustained out-of-phase oscillations. If $\tau_L > \tau_R$, then it is possible that both cells will lie in the silent phase on the left branch of C_0 at the same time. If there is no fixed point of this branch then the leading cell will be able to jump up. At this time, the trailing cell will approach the left branch of C_A and remain there until the leading cell jumps down. The trailing cell may then either jump up, if it lies below the left knee of C_0, or it may jump back to the left branch of C_0. In this latter case, it will eventually jump up when it reaches the left knee of C_0.

REMARK 5.12. Wang and Rinzel [**58**] distinguish between "escape" and "release" in producing out-of-phase oscillations. In the proof of the preceding theorem, the silent cell can only jump up to the active phase once the active cell jumps down and releases the silent cell from inhibition. This is referred to as the release mechanism and is often referred to as *postinhibitory rebound* [**12**]. To describe the escape mechanism, suppose that each cell is oscillatory for fixed levels of synaptic input. Moreover, one cell is active and the other is inactive. The inactive cell will then be able to escape the silent phase from the left knee of its cubic, despite the inhibition it receives from the active cell. Note that when the silent cell jumps up, it inhibits the active cell. This lowers the cubic of the active cell, so it may be forced to jump down before reaching a right knee.

6. Excitatory–Inhibitory Networks

6.1. Introduction. We now consider the network illustrated in Figure 9. This is composed of two populations of cells; these are labeled as E-cells and J-cells. We assume here that all of the E-cells are identical, as well as all of the J-cells; however, the E- and J-cells may differ from each other. Each E-cell sends excitatory synaptic input to some subpopulation of J-cells, while each J-cell sends inhibitory synaptic input to some subpopulation of E-cells. The J-cells may also receive inhibitory coupling from other J-cells. This type of network arises as models in several important neuronal systems, some of which are discussed later.

Excitatory–inhibitory networks can exhibit a rich structure of firing patterns. In a *synchronous oscillation*, the $E-$cells are completely synchronized with each other, as are the J-cells. Note that the E-cells are not necessarily synchronized with the J-cells, since we are not assuming that these cells are identical. In a *clustered oscillation*, each cell population forms subpopulations or clusters; the cells within a cluster fire synchronously but cells from distinct clusters act out of synchrony from each other. These types of networks can also exhibit propagating wave-like activity. A given network may display a wide variety of different sorts of waves, depending on network parameters or structure. For example, numerical simulation in [**37**] demonstrate that this type of network can exhibit at least two general classes of waves; these are referred to as *smooth waves* and *lurching waves*. A smooth wave corresponds to a fixed profile moving with constant velocity, much like a single action potential traveling down a nerve axon. If we consider a continuous model (as in [**56**]), then a solution vector $U(x,t)$ is a smooth wave if $U(x,t) = U(x - \mu t)$ for some constant μ corresponding to the wave speed. Note that if we keep track of where the leading edge of the membrane potential is some constant, say $v(x,t) = \theta$, then this defines a straight line of the form $x - \mu t =$ constant. A lurching wave does not propagate with constant velocity. Instead, the propagating wave recruits groups or clusters of cells in discrete steps. The leading edge of active cells inhibits some cluster of cells ahead of it; the size of this cluster depends on the synaptic footprint. The inhibited cells must wait until the active cells fall back to the silent phase, and then for inhibition to wear off, until the inhibited cells are able to jump up to the active phase. The firing of the cells within each cluster is nearly synchronous. Finally, these networks can exhibit chaotic firing patterns in which there is very little correlation between the firing of different neurons.

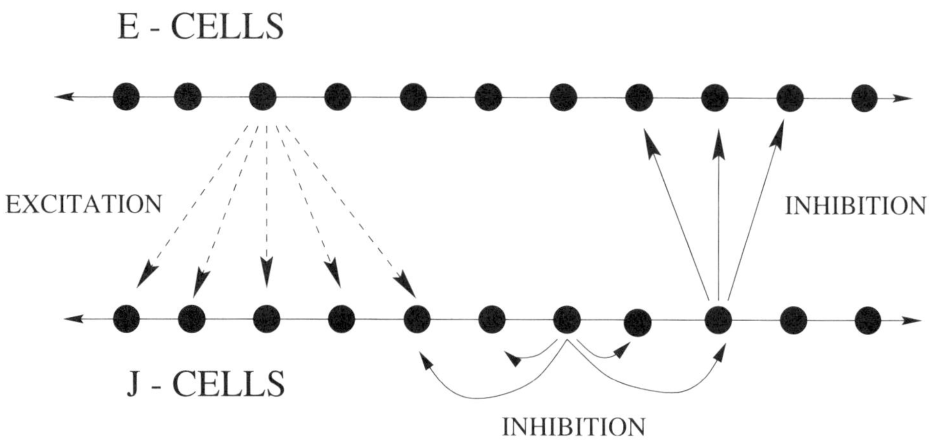

FIGURE 9. Excitatory-inhibitory network.

Each of the firing patterns described above arises in important applications. For example, during the spindle sleep rhythm, thalamic cells display clustered oscillations. More synchronized oscillations may correspond to a pathological epileptiform rhythm. Experiments have demonstrated that certain neurons within the basal ganglia display a chaotic rhythm during normal states; however, these neurons display a higher degree of correlation during pathological parkinsonian states. The firing patterns may resemble a clustered oscillation. More details of the applications are given in the next section.

6.2. Applications. Here we describe specific neuronal systems which have been modeled and analyzed using the methods described above. While we do not present the detailed equations, we note that each of these systems is modeled as an excitatory–inhibitory network and exhibits activity patterns of the types described in the previous section.

Image Segmentation: The problem of image segmentation is concerned with grouping similar sensory features of a perceived scene into coherent objects. Theoretical investigations of the brain and feature binding point to the mechanism of temporal correlation as a representational framework for image segmentation. The temporal correlation theory asserts that an object is represented by the correlated firing activity of the scattered neurons coding different features of an object. A natural way of encoding temporal correlation is to use neural oscillators, whereby each oscillator encodes some feature of an object. Each object is then represented by a group or block of oscillators that show synchrony of the oscillations, and different objects are represented by different groups whose oscillations are desynchronized from each other. This form of temporal or oscillatory correlation has received direct experimental support from cell recordings in the visual cortex and other brain regions. These experiments have triggered much interest in simulating the experimental results and revealing its underlying mechanisms.

In [**55**] a model for image segmentation based on locally excitatory and globally inhibitory oscillator networks is proposed and analyzed. In this model, the phases of oscillators encodes the binding of pixels. Rigorous mathematical analysis and extensive computational studies demonstrate the network's ability to both synchronize the oscillators within a particular block corresponding to some object and desynchronize the oscillators within different blocks corresponding to different objects.

Thalamic Sleep Rhythms: Biophysical models for the spindle and delta sleep rhythm are considered, in [**13, 51**], for example. This network consists of three types of cells; two of these lie in the thalamus while the other in the cortex. Each of these cell populations display different firing patterns during each stage of sleep and each of these patterns change from one stage of sleep to the other. The focus of [**51**] is to understand how the intrinsic and synaptic properties of these different cells interact to produce the different rhythms. It is demonstrated that a change in the conductance of one kind of cell effectively rewires the thalamocortical circuit, leading to the transition from the spindling to the delta rhythm. This study illustrates the different roles excitatory and inhibitory synaptic coupling can play in producing a particular rhythm. A geometric model for the spindle rhythm is developed and analyzed in [**38, 39**].

The Basal Ganglia: The basal ganglia comprise several nuclei which are thought to play an important role in the control of movement; it is also thought to be involved in cognition, motivation and emotion. Dysfunction of the basal ganglia is associated with movement disorders such as Parkinson's disease and Huntington's disease [**10**]. Structures within the basal ganglia are, in fact, the target of therapeutic surgical procedures including pallidotomy, lesioning of the subthalamic nucleus, and deep brain stimulation.

Numerous experiments have demonstrated that neurons within the basal ganglia display a rich variety of dynamic behavior; moreover, patterns of neuronal activity, both spatial and temporal, are different between a normal and a pathological state. For example, during normal resting conditions, certain neurons within the basal ganglia typically discharge in a tonic and irregular mode and there is little correlation in their spiking activity. However, during a Parkinsonian state, these neurons display a higher degree of rhythmic bursting activity and are less selective to specific movements. Recent experiments have also demonstrated that resting tremor is associated with synchronous activity among these neurons [**3**].

In [**54**] we develop and analyze a conductance-based network model for interactions of neurons in the external segment of the globus pallidus and the subthalamic nucleus of the basal ganglia. We found that the network robustly generates several types of activity patterns including two classes of synchronous rhythms, propagating wave-like activity and chaotic dynamics.

References

[1] W.B. Adams and J.A. Benson. The generation and modulation of endogenous rhythmicity in the *aplysia* bursting pacemaker neurone R15. *Prog. Biophys. Mol. Biol.*, **46**:1–49, 1985.

[2] S.M. Baer, T. Erneux, and J. Rinzel. The slow passage through a Hopf bifurcation: delay, memory effects, and resonances. *SIAM J. Appl. Math.*, **49**:55–71, 1989.

[3] H. Bergman, A. Feingold, A. Nini, A. Raz, H. Slovin, M. Abeles, and E. Vaadia. Physiological aspects of information processing in the basal ganglia of normal and parkinsonian primates. *Trends Neurosci.*, **21**:32–38, 1998.

[4] R.J. Butera, J. Rinzel, and J.C. Smith. Models of respiratory rhythm generation in the pre-Botzinger complex: I. Bursting pacemaker model. *J. Neurophysiology*, **82**:382–397, 1999.

[5] R.J. Butera, J. Rinzel, and J.C. Smith. Models of respiratory rhythm generation in the pre-Botzinger complex: II. Populations of coupled pacemaker neurons. *J. Neurophysiology*, **82**:398–415, 1999.

[6] T.R Chay and J. Keizer. Minimal model for membrane oscillations in the pancreatic β-cell. *Biophys. J.*, **42**:181–190, 1983.

[7] T.R. Chay and J. Rinzel. Bursting, beating, and chaos in an excitable membrane model. *Biophys. J.*, **47**:357–366, 1985.

[8] L.D. Cook, L.S. Satin, and W.F. Hopkins. Pancreatic β-cells are bursting, but how? *Trends Neurosci.*, **14**:411–414, 1991.

[9] G. de Vries and R.M. Miura. Analysis of a class of models of bursting electrical activity in pancreatic β-cells. *SIAM J. Appl. Math.*, **58**:607–635, 1998.

[10] M.R. DeLong. Activity of pallidal neurons during movement. *J. Neurophysiol.*, **34**:414–427, 1971.

[11] A.Destexhe, D.A. McCormick, and T.J. Sejnowksi. A model for 8-10 Hz spindling in interconnected thalamic relay and reticularis neurons. *Biophys. J.*, **65**:2474–2478, 1993.

[12] W. Friesen. Reciprocal inhibition, a mechanism underlying oscillatory animal movements. *Neurosci. Behavior*, **18**:547–553, 1994.

[13] D. Golomb, X.-J. Wang, and J. Rinzel. Synchronization properties of spindle oscillations in a thalamic reticular nucleus model. *J. Neurophysiol.*, **72**:1109–1126, 1994.

[14] J.L. Hindmarsh and R.M. Rose. A model of neuronal bursting using three coupled first order differential equations. *Proc. R. Soc. Lond., Ser. B*, **221**:87–102, 1984.

[15] A.L. Hodgkin and A.F. Huxley. A quantitative description of membrane current and its application to conduction and excitation in a nerve. *Journal of Physiology*, **117**:165–181, 1952.

[16] F.C. Hoppensteadt and E.M. Izhikevich. *Weakly Connected Neural Networks.* Springer-Verlag, New York, Berlin, and Heidelberg, 1997.

[17] E.M. Izhikevich. Neural excitability, spiking, and bursting. *International Journal of Bifurcation and Chaos*, **10**, 2000.

[18] D. Johnston and S. M-S. Wu. *Foundations of Cellular Neurophysiology.* MIT Press, Cambridge, Ma. and London, 1997.

[19] K. Kandler and L.C. Katz. Neuronal coupling and uncoupling in developing nervous systems. *Current Opin. Neurobiol.*, **5**:98–105, 1995.

[20] E.R. Kandel, J.H. Schwartz, and T.M. Jessell. *Principles of Neural Science.* Appleton & Lange, Norwalk, Conn., 1991.

[21] J. Keener and J. Sneyd. *Mathematical Physiology.* Springer, New York, Berlin, Heidelberg, 1998.

[22] N. Kopell and B. Ermentrout. Mechanisms of phase-locking and frequency control in pairs of coupled neural oscillators. In B. Fiedler, G. Iooss, and N. Kopell, editors, *Handbook of Dynamical Systems, vol. 3: Towards Applications.* Elsevier. To appear.

[23] N. Kopell and G.B. Ermentrout. Subcellular oscillations and bursting. *Math. Biosci.*, **78**:265–291, 1986.

[24] N. Kopell and D. Somers. Anti-phase solutions in relaxation oscillators coupled through excitatory synapses. *J. Math. Biol.*, **33**:261–280, 1995.

[25] E. Lee and D. Terman. Uniqueness and stability of periodic bursting solutions. To appear in *Journal of Differential Equations.*

[26] R.R. Llinas, T. Grace, and Y. Yarom. *In vitro* neurons in mammalian cortical layer 4 exhibit intrinsic oscillatory activity in the 10- to 50 Hz frequency range. *Proc. Natl. Acad. Sci. USA*, **88**:897–901, 1991.

[27] T. LoFaro and N. Kopell. Timing regulation in a network reduced from voltage-gated equations to a one-dimensional map. *J. Math. Biol.* To appear.

[28] C. Morris and H. Lecar. Voltage oscillations in the barnacle giant muscle fiber. *Biophys. J.*, **35**:193–213, 1981.

[29] C.A. Del Negro, C.-F. Hsiao, S.H. Chandler, and A. Garfinkel. Evidence for a novel bursting mechanism in rodent trigeminal neurons. *Biophys. J.*, **75**:174–182, 1998.

[30] A. Nejshtadt. Asymptotic investigation of the loss of stability by an equilibrium as a pair of eigenvalues slowly cross the imaginary axis. *Usp. Mat. Nauk*, **40**:190–191, 1985.

[31] M. Pernarowski. Fast subsystem bifurcations in a slowly varying Lienard system exhibiting bursting. *SIAM J. Appl. Math.*, **54**:814–832, 1994.

[32] P. Pinsky and J. Rinzel. Intrinsic and network rhythmogenesis in a reduced Traub model for CA3 neurons. *J. Comput. Neurosci.*, **1**:39–60, 1994.

[33] R.E. Plant. Bifurcation and resonance in a model for bursting nerve cells. *J. Math. Biol.*, **11**:15–32, 1981.

[34] J. Rinzel. A formal classification of bursting mechanisms in excitable systems. In A.M. Gleason, editor, *Proceedings of the International Congress of Mathematicians*, pages 1578–1594. American Mathematical Society, Providence, RI, 1987.

[35] J. Rinzel and G.B. Ermentrout. Analysis of neural excitability and oscillations. In C.Koch and I. Segev, editors, *Methods in Neuronal Modeling: From Ions to Networks*, pages 251–291. The MIT Press, Cambridge, MA, second edition, 1998.

[36] J. Rinzel and Y.S. Lee. Dissection of a model for neuronal parabolic bursting. *J. Math. Biol.*, **25**:653–657, 1987.

[37] J. Rinzel, D. Terman, X.-J. Wang, and B. Ermentrout. Propagating activity patterns in large-scale inhibitory neuronal networks. *Science*, **279**:1351–1355, 1998.

[38] J. Rubin and D. Terman. Geometric singular perturbation analysis of neuronal dynamics. In B. Fiedler, G. Iooss, and N. Kopell, editors, *Handbook of Dynamical Systems, vol. 3: Towards Applications*. Elsevier. To appear.

[39] J. Rubin and D. Terman. Analysis of clustered firing patterns in synaptically coupled networks of oscillators. *J. Math. Biol.*, **41**:513–545, 2000.

[40] J. Rubin and D. Terman. Geometric analysis of population rhythms in synaptically coupled neuronal networks. *Neural Comput.*, **12**:597–645, 2000.

[41] A. Sherman. Anti-phase, asymmetric and aperiodic oscillations in excitable cells – I. coupled bursters. *Bull. Math. Biol.*, **56**:811–835, 1994.

[42] A. Sherman. Contributions of modeling to understanding stimulus-secretion coupling in pancreatic β-cells. *American. J. Physiology*, **271**:547–559, 1996. p

[43] P. Smolen, J. Rinzel, and A. Sherman. Why pancreatic islets burst but single β-cells do not: The heterogeneity hypothesis. *Biophys. J.*, **64**:1668–1680, 1993.

[44] P. Smolen, D. Terman, and J. Rinzel. Properties of a bursting model with two slow inhibitory variables. *SIAM J. Appl. Math.*, **53**:861–892, 1993.

[45] D. Somers and N. Kopell. Rapid synchronization through fast threshold modulation. *Biol. Cybern.*, **68**:393–407, 1993.

[46] D. Somers and N. Kopell. Waves and synchrony in networks of oscillators of relaxation and non- relaxation type. *Physica D*, **89**:169–183, 1995.

[47] C. Soto-Trevino, N. Kopell, and D. Watson. Parabolic bursting revisited. *J. Math. Biol.*, **35**:114–128, 1996.

[48] J. Tabak, W. Senn, M.J. O'Donovan, and J. Rinzel. Comparison of two models for pattern generation based on synaptic depression. *Neurocomputing*, **20-27**:551–556, 1999.

[49] D. Terman. Chaotic spikes arising from a model for bursting in excitable membranes. *SIAM J. Appl. Math.*, **51**:1418–1450, 1991.

[50] D. Terman. The transition from bursting to continuous spiking in an excitable membrane model. *J. Nonlinear Sci.*, **2**:133–182, 1992.

[51] D. Terman, A. Bose, and N. Kopell. Functional reorganization in thalamocortical networks: Transition between spindling and delta sleep rhythms. *Proc. Natl. Acad. Sci. USA*, **93**:15417–15422, 1996.

[52] D. Terman, N. Kopell, and A. Bose. Dynamics of two mutually coupled inhibitory neurons. *Physica D*, **117**:241–275, 1998.

[53] D. Terman and E. Lee. Partial synchronization in a network of neural oscillators. *SIAM J. Appl. Math.*, **57**:252–293, 1997.

[54] D. Terman, J. Rubin, A.C. Yew, and C.J. Wilson. Synchronous rhythms in a model for the indirect pathway of the basal ganglia. *Neurocomputing*. To appear.

[55] D. Terman and D. L. Wang. Global competition and local cooperation in a network of neural oscillators. *Physica D*, **81**:148–176, 1995.

[56] D.H. Terman, G.B. Ermentrout, and A.C. Yew. Propagating activity patterns in thalamic neuronal networks. *SIAM J. Appl. Math.*, **61**:1578–1604, 2001.

[57] X.-J. Wang. Ionic basis for intrinsic-40 Hz neuronal oscillations. *Neuroreport*, **5**:221–224, 1993.

[58] X.-J. Wang and J. Rinzel. Spindle rhythmicity in the reticularis thalamic nucleus: synchronization among mutually inhibitory neurons. *Neuroscience*, **53**:899–904, 1993.

[59] X.-J. Wang and J. Rinzel. Oscillatory and bursting properties of neurons. In M.A. Arbib, editor, *The Handbook of Brain Theory and Neural Networks*, pages 686–691. The MIT Press, Cambridge, London, 1995.

DEPARTMENT OF MATHEMATICS, OHIO STATE UNIVERSITY, COLUMBUS, OHIO 43210
E-mail address: terman@math.ohio-state.edu

Proceedings of Symposia in Applied Mathematics
Volume **59**, 2002

Mathematics in Visual Neuroscience: The Retina

Daniel Tranchina

ABSTRACT. An introduction to retinal physiology is followed by a discussion of methods by which visual stimuli and neural responses can be described quantitatively. The concept of contrast and its central role in vision is discussed. The notion of a linear system is introduced, and its relevance to information processing in the retina is discussed. Two equivalent methods describing the input-output properties of a linear system are explained, the impulse response method and the transfer function method. Two generic linear systems relevant to visual physiology are analyzed, one involving a series arrangement of transducers, and the other involving a feedback arrangement. A common mechanism in the visual system of adjusting input-output properties through modulated feedback gain is illustrated. A linear description of the response properties of the retina's output neurons, the retinal ganglion cells, is given. The ganglion cell receptive field has an antagonistic center-surround organization. The spatial sensitivity profile for the retinal ganglion cells is sometimes described as a "Mexican hat" profile. The Mach band illusion is introduced, and the possible role of ganglion cells in this illusion is discussed. The neural basis for the ganglion cell receptive field structure is described qualitatively. A mathematical model for retinal ganglion cells is developed. A key component is a continuum model for the cone-horizontal cells network. Cones and horizontal cells interact reciprocally in a negative feedback loop. The model equations involve a pair of coupled partial differential equations. The equations are solved by Fourier methods. The emergence of a Mexican hat profile at the level of the bipolar cell which receives convergent input from numerous cones is demonstrated.

1. Introduction

Visual physiology is particularly amenable to mathematical description and mathematical analysis. The input to the visual system can be controlled and described with a great deal of precision by the vision neuroscientist or psychophysicist. The input is a time-varying pattern of light falling on the retina. A complete description is provided by $L(x, y, t, \lambda)$ which gives the number of photons per unit area per unit wavelength (λ) falling on the retina at each position (x, y) [1] and at each point in time, t. The visual system is quite accessible for electrophysiological

1991 *Mathematics Subject Classification.* 92C05, 92C20, 92C30.

Key words and phrases. Visual processing, retinal physiology, light adaptation, linear systems.

[1] In visual physiology it is customary to measure length in units of deg visual angle.

measurement of neural responses. Recordings can be made from the retina, optic nerve, lateral geniculate nucleus of the thalamus, and numerous visual cortical areas. In fact, even without any electrical recordings, a great deal of quantitative information about vision can be gathered by simply asking a human subject to report judgments in psychophysical experiments. As a consequence, there is a rich history of mathematics in visual neuroscience.

In this chapter, I shall review some applications of mathematics in vision, particularly in the area of retinal physiology. The mathematical tools will include differential equations and Fourier transforms. An excellent introduction to retinal physiology is provided by [**11**], which reviews a great deal of recent research. Another recent book by Wandell [**17**] gives a good introduction to quantitative techniques and mathematical modeling in visual neuroscience. A foundation for the approach in this chapter is the quantitative description of the interplay of excitation and inhibition in the the horseshoe crab retina given by Hartline and Ratliff [**2**]. The horseshoe crab is an invertebrate whose retina is much simpler than any vertebrate retina [**9**]. However, some key organizational principles of the horseshoe crab retina apply to the vertebrate retina as well. Thus, the mathematical techniques presented here are very much like the linear systems analysis methods used many years ago in modeling the retina of the horseshoe crab [**1**]. Numerous outstanding problems will require the application of methods for nonlinear systems analysis [**12**], [**16**].

Electrophysiological experiments often consist of measuring the electrical activity of individual neurons. In general, one finds that a particular neuron "responds" to light falling within a more-or-less well-defined area of the retina and that there is a distinguished point where response sensitivity is greatest. This area is called the *receptive field* of the neuron, and the point of peak sensitivity is called the receptive field center. Visual neurons give responses to a perturbation from steady illumination that can outlast this perturbation by tens or hundreds of milliseconds. However, the state of a neuron at time t is essentially independent of the stimulus previous to $t - T$, for large enough T. As a consequence, the neurons are said to have finite memory. It is often convenient to label a neuron by the location of its receptive field center, say (x_0, y_0), and then to define its activity or response as a function of time $r(x_0, y_0, t)$. From the point of view of a mathematician, there is a mapping of the stimulus for all times previous to time t to the response of the neuron at time t, $r(x_0, y_0, t)$. In other words, r is a functional of the light stimulus L.

Much of what vision scientists do can be regarded as efforts to characterize the functionals that link visual inputs to responses and then to account for them in terms of underlying neural and biophysical mechanisms. This chapter will focus on both these efforts. But first, a more physiological overview of vision is in order. My comments will be confined to visual information processing in the retina.

2. The Retina

The retina acts as a transducer; it transforms the visual image into an electrical signal which is sent to higher brain areas via the optic nerve. The optic nerve consists of roughly 10^6 axons emanating from the retinal ganglion cells, the output neurons of the retina. Each optic nerve fiber (axon) carries its own signal in the form of discrete electrical impulses that propagate without decrement to target areas. These impulses are called action potentials, which can be measured by inserting a micro-electrode into the axon in order to record the voltage difference across the axonal membrane. The action potential is then seen as a brief (on the order of 1 ms) transient reversal of the voltage difference across the membrane. The same signals can be recorded by inserting micro-electrodes into the cell bodies of retinal ganglion cells. The rate of action potentials, often called spikes, is modest. In the presence of spatially uniform illumination, retinal ganglion cells will "fire" spikes at a rate of roughly 20–50 per second. This rate depends surprisingly little on the level of the steady illumination. The visual system is designed to detect spatial and temporal contrast in visual scenes rather than to encode absolute light intensities. When the light is modulated, the rate of spikes varies between 0 and roughly 200 spikes per second. The spike train might be thought of as a digital system of encoding. If one divides time up into brief enough intervals, then each interval either contains or doesn't contain a spike.

The spike train in response to a given visual stimulus is a stochastic point process; both the total number of spikes and the arrival times of these spikes vary from stimulus presentation to stimulus presentation. Stochastic information processing is the rule throughout the visual system. Consequently, when rate is used as measure of response, it is usually a derived quantity which is obtained by presenting the same stimulus repeatedly and collecting spikes that are put into bins phased to the stimulus cycle; the rate is then given by dividing the total number of spikes in each bin by the total time represented by each bin. In other words, the notion of spike rate is only well-defined in the context of a signal-averaged response. The method by which stochastic spike trains are decoded by higher areas in the brain is not at all well understood.

2.1. Retinal circuitry. Complex information processing within the retina precedes the output spike train. The primate retina [**11**] is a layered network of neurons, roughly 500 microns (1 micron= 10^{-6} m) thick. It consists of three cell body or nuclear layers separated by two synaptic or plexiform layers (Figure 1). The outer nuclear layer contains the cell bodies of rod and cone photoreceptors. Rods and cones subserve night and day vision, respectively. There are three types of cones with different peak spectral sensitivities at long, medium and short wavelengths. They mediate color vision. The inner nuclear layer contains the cell bodies of horizontal cells, bipolar cells and amacrine cells. The third cell body layer contains the retinal ganglion cells (output neurons). There is a straight-through pathway for information flow from photoreceptors to bipolar cells to retinal ganglion cells. There are two lateral pathways provided by horizontal cells in the outer plexiform layer and provided by amacrine cells in the inner plexiform layer.

In the outer plexiform layer, photoreceptors make synapses onto bipolar and horizontal cells, and some horizontal cells make feedback synapses onto cones. In the inner nuclear layer bipolar cells make feed-forward synapses onto amacrine cells and ganglion cells. Amacrine cells make feed-forward synapses onto ganglion and amacrine cells, and make feedback synapses onto bipolar cells.

Within the retina, information is largely encoded in an analog manner; photoreceptors, bipolar cells and amacrine cells respond to visual stimuli with smooth and slow modulation of their membrane voltages. Thus, the information is encoded in the membrane potential itself rather than in the rate or times of impulses.

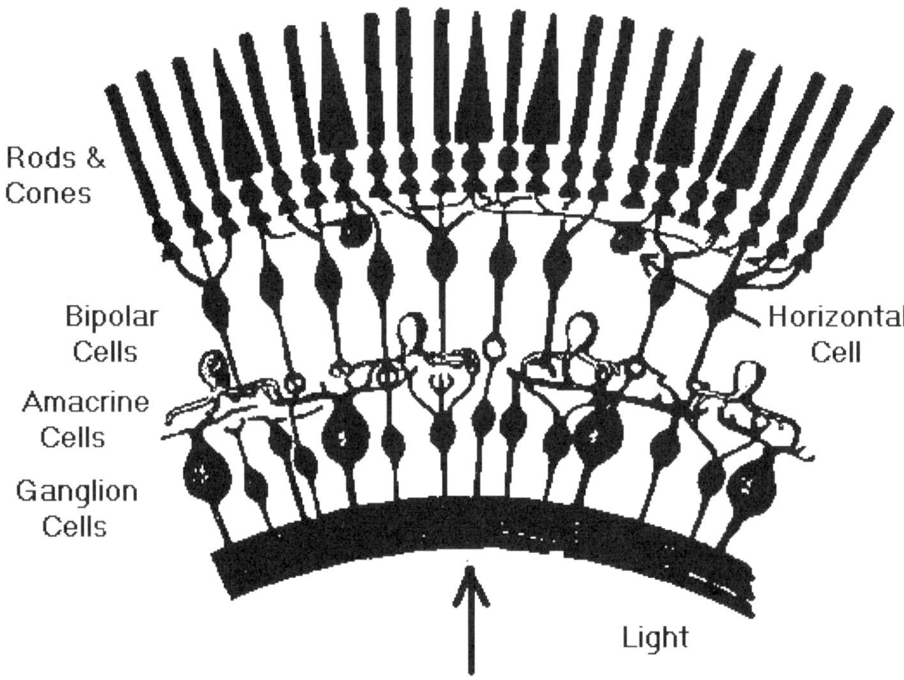

FIGURE 1. Schematic diagram of the retina, consisting of three cell body or nuclear layers separated by two synaptic or plexiform layers. This diagram is a modified version of the file schem.jpeg on the web site http://webvision.med.utah.edu, and is reproduced here with permission.

3. Quantifying the Light Stimulus

For purposes of simplifying the discussion, color vision will be ignored in the following. The reader is asked to think of light in terms of photons per unit area per unit time without regard to the spectral composition of that light. It is sensible to do so, because the kinds of vision experiments we will be dealing with here involve visual stimuli consisting of modulations in the intensity of light with fixed spectral composition.

Physiological stimuli consist largely of moderate perturbations of light around some mean level. It is useful to characterize these perturbations in terms of contrast. To do so, the light stimulus is written as

$$\begin{aligned} L(\vec{x},t) &= L_0 + L_1(\vec{x},t) \\ &= L_0\Big(1 + \frac{L_1(\vec{x},t)}{L_0}\Big), \end{aligned} \tag{3.1}$$

where L_0 is the mean light level, and L_1 is the perturbation about the mean. The term $\frac{L_1(\vec{x},t)}{L_0}$ is called *contrast*, which we shall denote by the term $s(\vec{x},t)$. The contrast is a signed quantity and it is independent of the level of illumination,

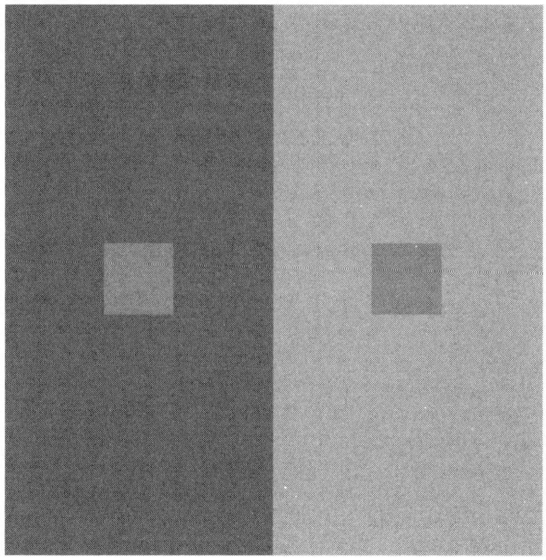

FIGURE 2. Demonstration that perceived brightness is determined by contrast rather than absolute light intensities. The intensity of reflected light is identical for the two square regions in this figure. Most subjects perceive the square on the left as brighter than the square on the right. The contrast of the square on the left is said to be positive because its physical intensity is greater than that of the local background region; the contrast of the square on the right is said to be negative, because its physical intensity is less than that of its local background region.

because both I_0 and I_1 scale linearly with the level of illumination. Thus, the contrast in a scene (not containing self luminant objects such as a television) is characteristic of the physical attributes of objects in the scene; it is independent of overall illumination, which can vary over many orders of magnitude over the course of a day. In fact, the ratio of the luminance of snow in sunlight to the luminance at rod threshold is 10^{11} [**11**]. Thus, the retina mediates vision over an eleven log-unit range of light levels; the range is even greater if we allow for extreme conditions which are occasionally encountered (e.g. looking accidentally directly into bright headlights).

The retina presents to higher brain areas a signal that is more closely related to contrast than to absolute light levels. This contrast coding has implications for visual perception, and one is demonstrated in Figure 2. The two square patches in this figure are of identical physical luminance. The local contrast in going from the background rectangular area to the square patch on the right is negative, and that patch is perceived as dark. The situation is just the opposite on the left.

4. Linear Information Processing

Photoreceptors, horizontal cells and bipolar cells respond approximately linearly to physiological stimuli. That is, the functional that relates $s(\vec{x}, t)$ to $r(\vec{x}_0, t)$

is approximately linear for each of these cell types. However, this linear functional depends on the operating point (background light level L_0). Linearity of the response means that the response to the sum of two stimuli $s_1(\vec{x},t)$ and $s_2(\vec{x},t)$ presented individually is given by the sum of the responses, $r_1(\vec{x}_0,t)$ and $r_2(\vec{x}_0,t)$, to the two individual stimuli. In addition, visual neurons behave in a time invariant (stationary) manner over time periods that are long on the time scale of response to a transient stimulus. That is, if the response to $s(\vec{x},t)$ is $r(\vec{x}_0,t)$, then the response to $s(\vec{x},t-\tau)$ is $r(\vec{x}_0,t-\tau)$; translating the stimulus on the time axis by τ simply translates the response by τ.

4.1. Impulse response function. The response properties of a linear, time-stationary neuron are completely determined by its *impulse response function*, $k(\vec{x},t;L_0)$. To say that the impulse response function is $k(\vec{x},t;L_0)$ means that the response to the stimulus [2] $s(\vec{x},t) = \delta(\vec{x}-\vec{x}_1,t)\,\mathrm{deg}^2\,\mathrm{s}$ is $k(\vec{x}_1-\vec{x}_0,t;L_0)$. In practice, a laboratory realization of a Dirac delta function stimulus is a flash of light confined to an area that is very small on the scale of the receptive field size and of a duration that is short on the time scale of the response.

As a consequence of linearity and translational invariance in time, the response to an arbitrary stimulus $s(\vec{x},t)$ is given by

$$r(\vec{x}_0,t) = \int d\vec{x} \int dt'\, s(\vec{x},t')k(\vec{x}-\vec{x}_0,t-t';L_0). \tag{4.1}$$

There is also a physical argument behind this convolution equation involving an approximation of the stimulus by discretizing in time and space. One makes the reasonable assumption that retinal area can be subdivided into pixels that are so small, and time can be divided into bins so short that only the total incremental number of photons in these pixels for each time bin matters in determining the response. That is, discretizing the stimulus more finely would hardly alter the response. In this regime, the stimulus in each pixel and in each time bin is replaced with a δ-function with amplitude given by the total number of photons in each pixel for each time bin. Then the superposition principle is used to compute the overall response. The approximation is better the more finely divided are time and space. In the limit, the summation gives the convolution integral.

4.2. Transfer function or frequency response. Linear response properties can also be characterized in the frequency domain. To do so, one considers a stimulus of the form

$$s(\vec{x},t) = m\,\mathrm{Re}\exp(-i2\pi[\vec{x}\cdot\vec{\xi}-ft]), \tag{4.2}$$

where Re denotes the real part. This stimulus is a drifting sinusoidal grating. A snapshot of this sort of stimulus is shown in Figure 3. At each instant in time, the illuminance varies as a sinusoidal function of position. This is a one-dimensional rectilinear stimulus; the illuminance is constant along lines in the retinal plane that are perpendicular to the ξ-vector, and the spatial frequency of sinusoidal variation in the direction of the ξ-vector is $|\vec{\xi}|$ (units of cycles/deg). At each fixed position, the illuminance varies with a sinusoidal time course with temporal frequency f (units Hz, i.e. cycles/s).

[2] If α is the number of photons per deg^2 per s in the background light level L_0 then this contrast stimulus corresponds to the incremental stimulus $L_1 = \alpha\,\delta(\vec{x}-\vec{x}_1,t)$; i. e. this stimulus delivers a total of α photons.

FIGURE 3. A sinusoidal grating stimulus. Contrast s in this image varies according to equation (4.2) with fixed t.

If we specialize (4.1) to (4.2) then

$$r(\vec{x}_0, t) = \mathrm{Re} \int d\vec{x} \int dt' \exp(-i2\pi[\vec{x} \cdot \vec{\xi} - ft'])\, k(\vec{x} - \vec{x}_0, t - t'; L_0). \tag{4.3}$$

For sake of simplicity, let us define the receptive field center $\vec{x}_0$ as the origin and drop the spatial argument of r. Then

$$r(t) = \mathrm{Re}\,[m\, e^{i2\pi ft} K(\vec{\xi}, f; L_0)], \tag{4.4}$$

where K is the Fourier transform or the spatiotemporal impulse response function. K is called the spatiotemporal transfer function or spatiotemporal frequency response. If we write K in complex exponential form, as $K = |K|e^{\theta}$, then

$$\begin{aligned} r(t) &= m\, \mathrm{Re}\, |K| e^{i2\pi ft + \theta} \\ &= |K| \cos(2\pi ft + \theta) \end{aligned} \tag{4.5}$$

Equation (4.5) means that the response to a drifting sinusoidal grating stimulus is a sinusoidal function of time at the drift frequency. Furthermore, the response per unit contrast, $r(t)/m$, has an amplitude and phase angle equal to the modulus and argument, respectively, of the spatiotemporal transfer function.

There are advantages to measuring $K(\vec{\xi}, f; L_0)$ instead of $k(\vec{x}, t; L_0)$ in a system that is not perfectly linear. Measurement of $k(\vec{x}, t; L_0)$ requires a flash stimulus of large enough magnitude to give a response that stands out from the noise, and such a stimulus could easily drive the system out of its (approximately) linear regime. This is much less of a problem with a sinusoidal stimulus. Moreover, nonlinear systems with finite memory will in general respond to a periodic stimulus with a periodic response at the input period. Consequently, the Fourier components of the response will be at integer multiples of the fundamental frequency of the stimulus. Therefore, nonlinearities in the response to a drifting sinusoidal grating would be immediately revealed by Fourier components of the response at integer multiples of the input frequency, f. If the response is only mildly nonlinear, then the fundamental Fourier component of the response (i.e. the component at the input frequency), obtained by taking the Fourier transform of the response, provides a good approximate measure of K. The Fourier transform operation also effectively filters out noise, which has power distributed across all frequencies and thus not so much at any one particular frequency.

There are also analytical advantages to representing input-output properties of linear networks in the frequency domain. I show below that a system comprised of a series arrangement of linear transducers with or without feedback has a simple transfer function that can be expressed as an algebraic combination of the transfer functions of individual transducers. It is often not possible to write down the corresponding impulse response function.

4.3. Series arrangement of transducers. For the sake of streamlining the notation, we will consider an input-output system in which the input and output are functions of time only. In the context of visual physiology, this corresponds to focusing on the response of a neuron to temporal modulation of spatially uniform illumination. Consider the series arrangement of transducers, represented by the boxes, in Figure 4, upper panel. This figure shows a circuit diagram in which each transducer is represented by a box. Each transducer is labeled with its impulse response function. The impulse response functions of transducers 1 and 2 are $k_1(t)$ and $k_2(t)$, respectively. The input to transducer 1 is $s(t)$, and its output is $x(t)$. The output of transducer 1, $x(t)$, is the input to transducer 2, and its output is $y(t)$.

In response to $s(t) = \delta(t)$, $x(t) = k_1(t)$ and $y(t) = (x * k_2)(t) = (k_1 * k_2)(t)$ Thus, the impulse response of the system as a whole (the system enclosed in a dotted rectangle, Figure 4, upper panel), $k(t)$, is given by $k(t) = (k_1 * k_2)(t)$. The general conclusion is that the impulse response of a concatenation of linear transducers is given by the convolution of the impulse responses of the individual transducers.

The corresponding transfer function or frequency response function can be found in two ways. One is by using the fact that the transfer function is the Fourier transform of the impulse response function, and the fact that the Fourier transform of the convolution of two functions is given by the product of the Fourier transforms. Therefore, $K(f) = K_1(f)K_2(f)$. The same result may be obtained by thinking about the system in the frequency domain as shown in Figure 4, lower panel. We simply use the fact that the response of a linear system to a complex exponential is that complex exponential multiplied by the transfer function. Thus, if $s(t) = e^{i2\pi ft}$, then $x(t) = K_1(f)e^{i2\pi ft}$, and $y(t) = K_2(f)(K_1(f)e^{i2\pi ft}) = [K_1(f)K_2(f)]e^{i2\pi ft}$. By inspection, the overall transfer function $K(f)$ is given by $K(f) = K_1(f)K_2(f)$.

4.4. Subtractive feedback arrangement of transducers. Feedback is common throughout the visual system. Therefore, it is worthwhile to explore the frequency response of the generic feedback system shown in Figure 5. This figure shows 3 transducers. The input is to a transducer with frequency response (transfer function) $A(f)$. The output of this transducer, $x(t)$, feeds into a summation point where it combines with a feedback signal, $z(t)$. The output from the summation point is called $y(t)$, where $y(t) = x(t) - z(t)$. The signal $y(t)$ feeds into a transducer with frequency response $B(f)$, and the output of this transducer is

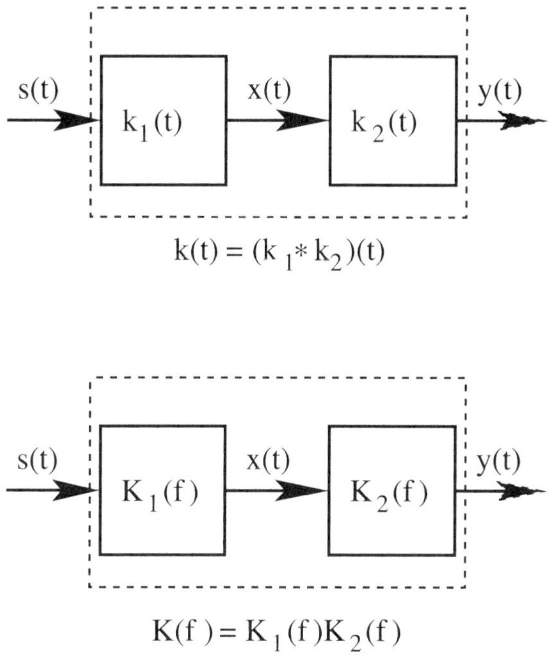

FIGURE 4. Series arrangement of two transducers. Each transducer is denoted by a box. In the upper panel, each transducer is labeled by its impulse response function, and in the lower panel each is labeled by its transfer (temporal frequency response) function.

the response (output) of the system. The output signal is tapped off and fed back to the summation point: the output, $r(t)$ feeds into a transducer with frequency response $C(f)$, and the output of this transducer is the feedback signal $z(t)$.

The transfer function of this system can be found by solving an algebraic problem. We consider sinusoidal input and write each signal in complex exponential form: $s(t) = e^{i2\pi ft}$; $x(t) = Xe^{i2\pi ft}$; etc. Then, by keeping track of the sinusoidal input and output of each transducer and by taking note that the signal $y(t) = x(t) - z(t)$, we can solve for the overall frequency response of the system:

$$\begin{aligned} X &= A(f) \\ Y &= X - Z \\ Z &= C(f)\,R \\ R &= B(f)\,Y \\ &= B(f)(X - C(f)\,R) \\ R &= \frac{A(f)B(f)}{1 + B(f)\,C(f)} \end{aligned} \tag{4.6}$$

The results above on series and feedback arrangements of transducers in which the input and output are functions of time only, generalizes to transducers in which the input and output are functions of time and space. We will use this fact below to

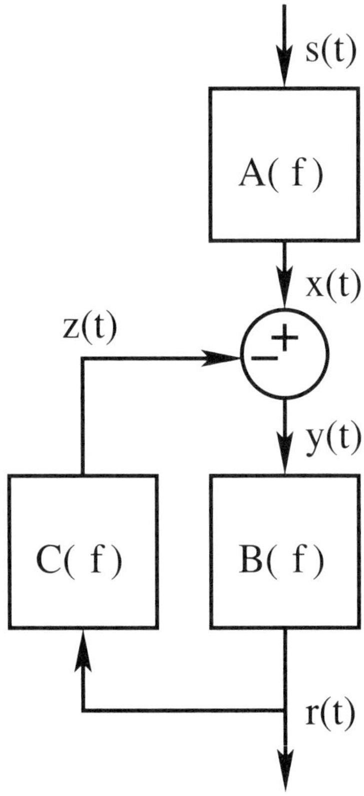

FIGURE 5. Subtractive feedback arrangement of transducers.

develop a model that accounts for the spatiotemporal response properties of retinal ganglion cells.

4.4.1. *Adjustment of gain and dynamics through modulated feedback.* A recurring theme in the visual system is the adjustment of gain, kinetics, and spatial aspects of the response properties of neurons through modulated feedback. Figure 6 (left panel) shows a simple subtractive feedback system (special case of that in Figure 5) that demonstrates this mechanism. The feed-forward pathway contains a transducer whose temporal frequency response and impulse response functions are $B(f) = 1/(1 + i2\pi f\tau)$ and $b(t) = H(t)e^{-t/\tau}/\tau$, respectively, where $H(t)$ is the Heaviside step function. The feedback pathway contains a transducer that provides no temporal filtering (its impulse response function is a Dirac delta function); it simply multiplies the feedback signal by a gain factor, g. The overall impulse response and frequency response functions of the system are

$$k(t) = H(t)\frac{1}{\tau}\exp\left(\frac{-t}{\frac{\tau}{1+g}}\right), \tag{4.7}$$

and

$$K(f) = \left(\frac{1}{1+g}\right)\left[\frac{1}{1+i2\pi f\left(\frac{\tau}{1+g}\right)}\right], \tag{4.8}$$

respectively. Note that the time constant of the exponential impulse response function (4.7) is given by $\tau/(1+g)$; it decreases as the feedback gain, g, increases. The area under the impulse response function, one measure of gain, is given by $1/(1+g)$. A family of temporal frequency (transfer) functions for different values of feedback gain g (0.1, 1, 10, 100) is shown in the right panel of Figure 6, where the uppermost curve is for the smallest value of g (0.1), and the lower-most curve is for the largest value of g (100). A characteristic feature of such a system is that the output gain at low frequencies decreases with increasing feedback gain g (asymptotically like $1/g$), while the gain at high frequencies is independent of g.

An important concept in linear systems theory is the *gain-bandwidth product.* One measure of bandwidth is the temporal frequency at which gain is reduced by a factor of $\sqrt{2}$ from its zero-frequency limit. This frequency, sometimes called the corner frequency, according to equation (4.8), is given by $(1+g)/(2\pi\tau)$, while the overall gain, as defined above, is given by $1/(1+g)$. Thus, the product of the gain and bandwidth is constant for this feedback system with adjustable feedback strength (gain). The conclusion is that response kinetics can be sped up at the expense of overall gain.

The qualitative behavior of this system is much like that of cone photoreceptors, in which feedback gain increases with increasing light level, through a nonlinear feedback mechanism. Consequently, the gain (sensitivity) of cones decreases with increasing light level, while the kinetics of the cone response speed up [**15**]. This phenomenon has been analyzed in the context of optimal filtering of photon noise [**8**].

5. Retinal Ganglion Cells.

The spatiotemporal impulse response function of retinal ganglion cells can be approximated by the empirical equation,

$$k(\vec{x},t) = \left[\frac{a_c}{2\pi\sigma_c^2}\exp\left(\frac{-|\vec{x}|^2}{2\sigma_c^2}\right) - \frac{a_s}{2\pi\sigma_s^2}\exp\left(\frac{-|\vec{x}|^2}{2\sigma_s^2}\right)\right]g(t) \tag{5.1}$$

This impulse response function is said to be spatiotemporal separable, because it is a product of a spatial function and a temporal function. The spatial function is a difference of Gaussians representation of spatial sensitivity [**10**]. Spatial profiles like this are sometimes called "Mexican hat" profiles. A plot of the spatial part of the impulse response function (cross-section through the origin) is shown in Figure 7, upper panel.

The ganglion cell spatiotemporal function corresponding to (5.1) is

$$K(\vec{\xi},f) = \left[a_c\exp\left(\frac{-\sigma_c^2|\vec{\xi}|^2}{2}\right) - a_s\exp\left(\frac{-\sigma_s^2|\vec{\xi}|^2}{2}\right)\right]G(f). \tag{5.2}$$

The spatial frequency part of the transfer function (5.2) is plotted in Figure 7, lower panel.

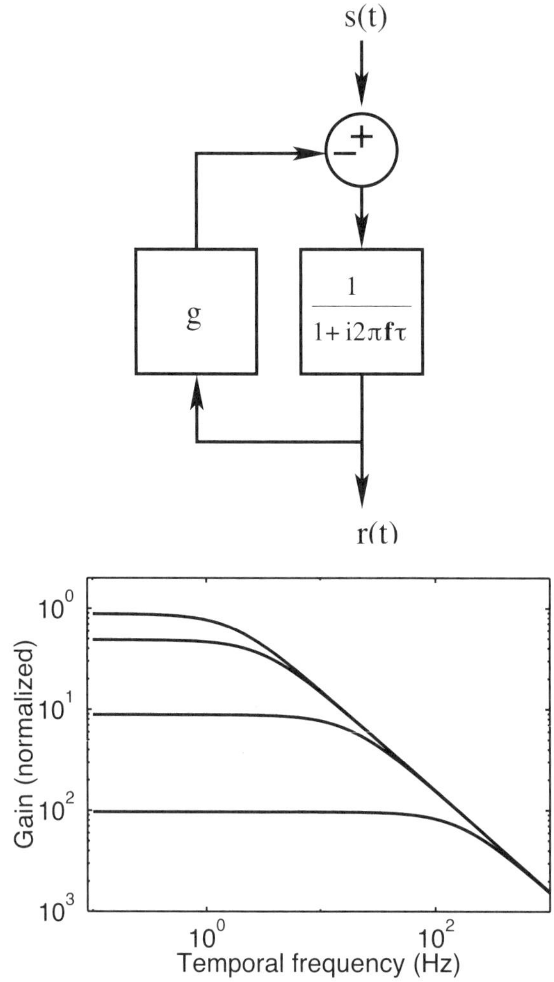

FIGURE 6. Adjustment of gain and kinetics through modulated feedback. Upper panel: the feed-forward pathway contains a transducer whose temporal frequency response and impulse response functions are $B(f) = 1/(1 + i2\pi f\tau)$ and $b(t) = H(t)e^{-t/\tau}/\tau$, respectively. Lower panel: a family of temporal frequency (transfer) functions for different values of feedback gain g (0.1, 1, 10, 100) where the uppermost curve is for the smallest value of g and the lowermost curve is for the largest value of g.

This sort of antagonistic center-surround receptive field organization, discovered by Kuffler [**6**], is thought to play a role in edge enhancement and edge detection in our visual system.

5.1. The Mach band illusion. Figure 8, upper panel, demonstrates a famous illusion known as the Mach band illusion (after Ernst Mach) [**9**]. This figure appears to consist of a uniform gray area on the left, a darker uniform gray area

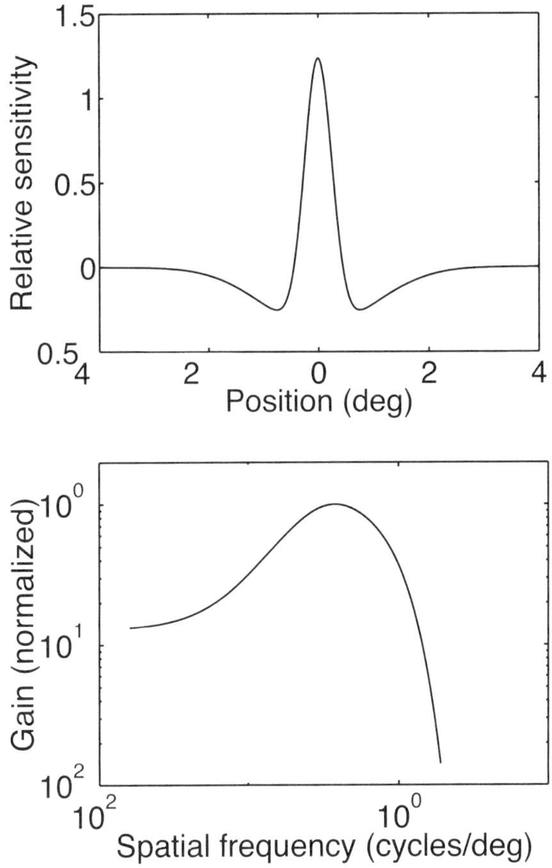

FIGURE 7. Retinal ganglion cell spatial sensitivity profile and corresponding spatial transfer function. Upper panel: spatial sensitivity profile for a retinal ganglion cell, the spatial term in equation (5.1). Lower panel: The spatial frequency counterpart, the spatial frequency term in equation (5.2).

on the right, separated by a transition region with a bright band just to the left of center, and a dark band just to the right of center. The actual luminance profile is a graduated edge plotted in the lower panel of Figure 8. There is, in fact, a constant gradient of illuminance separating the two uniform gray areas.

Some basis for implicating retinal ganglion cells in this illusion is demonstrated in Figure 9. The upper panel, plots the illuminance of a steady stimulus that varies in only the x-direction, a graded edge like that in Figure 8. The steady-state response rate of retinal ganglion cells, with receptive fields described by equation (5.1), is plotted as a function of position in the lower panel. Note the enhanced activity just to the left of the edge and the depressed activity just to the right of the edge. For purposes of illustration, the spatial sensitivity profile in Figure 7 is like those found in the cat retina. The antagonistic surround of ganglion cells in the primate retina is not quite as strong as that shown in Figure 7. It might be that

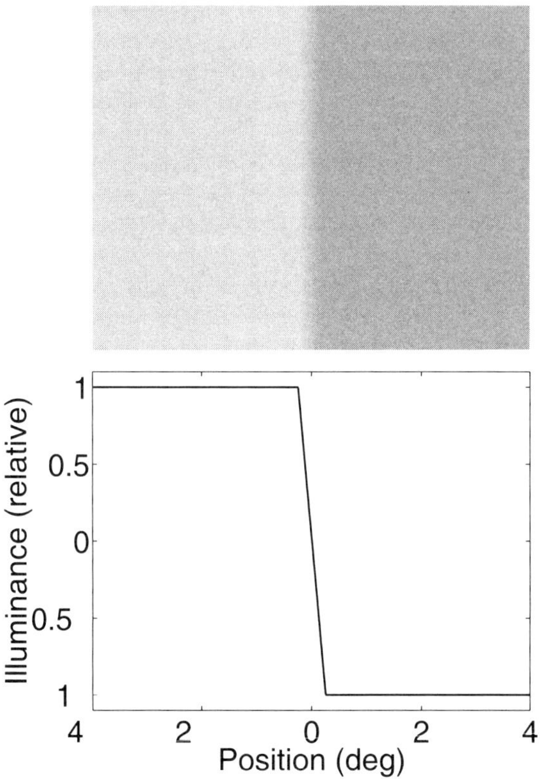

FIGURE 8. The Mach bands illusion. The upper panel of this figure appears to consist of a uniform gray area on the left, a darker uniform gray area on the right, separated by a transition region with a bright band just to the left of center, and a dark band just to the right of center. The actual luminance profile is a graduated edge plotted in the lower panel There is, in fact, a constant gradient of illuminance separating the two uniform gray areas.

neurons in primary visual cortex, which also have an antagonistic center-surround receptive field organization, play a more important role than retinal ganglion cells in the Mach band illusion.

6. Neural Basis for Ganglion Cell Receptive Field Structure

6.1. Continuum model for cones and horizontal cells. The foundation for the antagonistic center-surround receptive field organization of retinal ganglion cells is laid in the outer plexiform layer by reciprocal interactions between cones and horizontal cells. The feed-forward pathway from cones to horizontal cells is via a classical sign-conserving synapse. Feedback is via an unusual electrical modulation of the calcium current in the cone synaptic terminal membrane and consequent modulation of transmitter release [**4**]. For simplicity, this feedback will be modeled as if it were a sign inverting synapse.

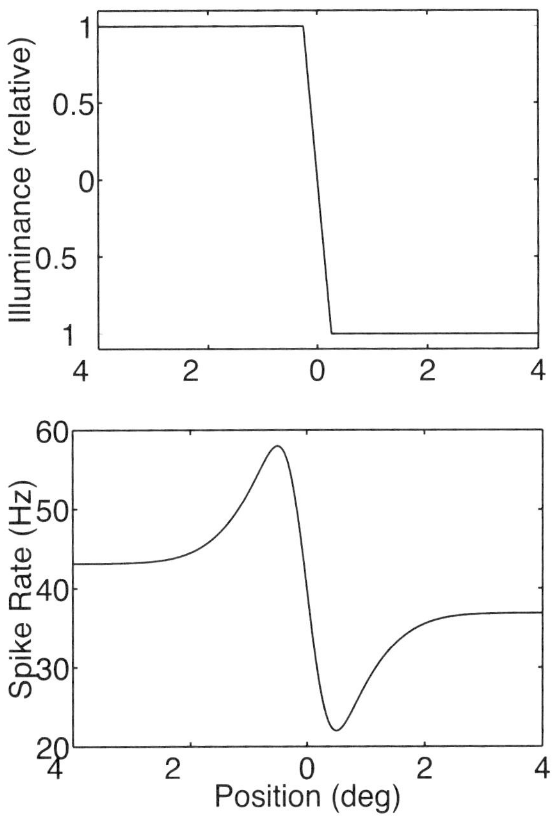

FIGURE 9. Possible electrophysiological analogue of the Mach band illusion. The upper panel, plots the illuminance of a steady (in time) stimulus that varies in only the x-direction. This profile corresponds to a graduated edge like that in the Mach band demonstration above (Figure 8, upper panel). The lower panel plots the steady-state response rate of retinal ganglion cells with spatiotemporal impulse response function described by equation (5.1). The enhanced activity just to the right of the edge and the depressed activity just to the left can be thought of as an electrophysiological analogue of the brightness perception when a subject views the upper panel in Figure 8.

In the peripheral retina, cones of like spectral sensitivity are electrically coupled to each other by gap junctions and thereby form a syncytium. The space scale for signal spread in the electrically coupled network of cones is large compared to the spacing between neighboring cones. Consequently, the discrete network of individual cells can be well-approximated by a continuum for the purposes of modeling the spread of electrical signals in the network. The cone network is regarded as a homogeneous leaky slab in which voltage varies in the $x - y$ plane, but not in the z-dimension, as the slab is very thin. The horizontal cells form a similar network with a space scale for signal spread which is about ten times larger

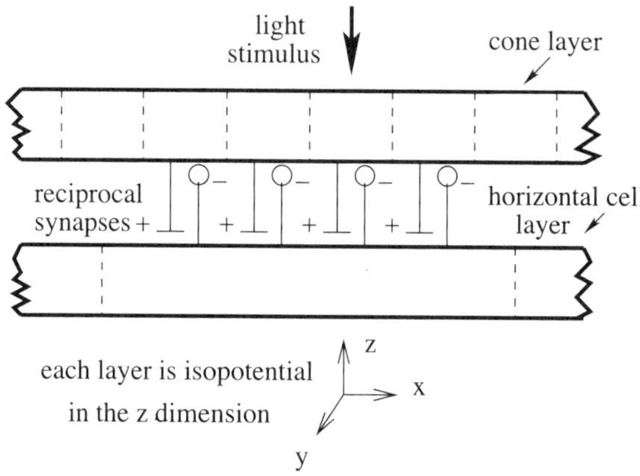

FIGURE 10. The cone-horizontal cell network. This is a schematic diagram illustrating a network of electrically coupled cones, delineated by dotted lines, in the upper layer of the figure, and a similar network of horizontal cells in the lower layer. Each network is modeled as a homogeneous slab. The two networks interact pointwise by reciprocal synapses. The feed-forward synapse between cones and horizontal cells is sign-conserving, and the feedback synapse from horizontal cells to cones is sign-inverting. The space scale for signal spread in the horizontal cell network is about ten times larger than the space scale for signal spread in the cone network.

than that in the cone network. A schematic diagram of the cone-horizontal cell network is shown in Figure 10. This model is similar to that of Krausz and Naka [**5**], and also similar to that of Tranchina et al. [**14**].

The equation describing the relationship between voltage and current in the cone and in the horizontal cell network is a two-dimensional analogue of the cable equation. To develop this equation, let us focus on the cone network. It is helpful to distinguish between any source currents and the resulting currents that flow within the cone network (slab) or across the cone (slab) membrane. The source currents are of two types. One is the *light-sensitive current.* This current is gated by an internal transmitter whose concentration is controlled by light through the phototransduction mechanism. The other is synaptic current from horizontal cell feedback synapses onto cones. Although there is, in general, a nonlinear relationship between membrane conductances and voltage, there is a regime of conductance values in which the modulation of conductance can be well-approximated by the injection of current from an equivalent current source. In this physiologically relevant regime, synaptic input can be modeled as current injection (source current).

In general, the membranes of neurons contain time- and voltage-dependent conductances. The sodium and potassium conductances in the giant axon of the squid, analyzed by Hodgkin and Huxley [**3**], are famous examples. Retinal neurons contain relatives of these conductances which respond dynamically to voltage changes across the cell membrane, but in general do not lead to regenerative currents that give action potentials. These conductances make the membranes of retinal neurons nonlinear. However, in the regime of physiological stimuli, the voltage modulations are small enough that the linearized equations describing the evolution of these conductances provide excellent approximations. In this regime, the time- and voltage-dependent conductances endow the membrane with a phenomenological or equivalent impedance.

With this background, we are ready to write the membrane voltage equation for the cone network. A circuit diagram for the cone-horizontal cell network is shown in Figure 11. The transfer functions of the transducers and their arrangement are based on the following arguments. The phototransduction mechanism in cones transforms the light into a current across the cone membrane (cone photocurrent). The transfer function for this transduction is $-P(f)$, where the minus sign reflects the polarity of the current. The photocurrent acts like a current source into the network of cones. The cones form a syncytium which behaves like a homogeneous slab of tissue. Current can spread laterally in this slab and it can also leak across the membrane of the slab. The photocurrent is combined algebraically (circle with plus and minus signs) with another source current from horizontal cell feedback. The net source current is transformed into cone membrane potential by the cone membrane, and the spatiotemporal transfer function for this transduction appears between the cone membrane current and cone membrane potential signals. At the synapse between cones and horizontal cells, a perturbation in the cone voltage leads to an injection of current into the horizontal cell network. The transfer function for this transduction is $F(f)$. The horizontal cells are electrically coupled, just like the cones, but the space scale for current spread is larger in the horizontal cell network. The feed-forward synaptic current is transformed into a perturbation in the horizontal cell membrane voltage by the horizontal cell membrane. Through the feedback synapse, a perturbation in the horizontal cell membrane voltage leads to a current injection into cone. The transfer function for this transduction is $B(f)$.

Let us denote the voltage across the cone membrane by $u(\vec{x}, t)$; the conductance for the spread of current within the cone network by σ_c, the total source current by $i_c^s(\vec{x}, t)$; the current across the equivalent cone membrane impedance by $i_c^Z(\vec{x}, t)$, and the equivalent impedance of the cone membrane by $Z_c(f)$. Applying the principle of current conservation in conjunction with Ohm's law gives

$$\sigma_c \nabla^2 u = i_c^Z - i_c^s \tag{6.1}$$

Now we can easily find the relationship between source current i_c^s, regarded as the input to cone, and cone membrane voltage u, regarded as the output. It will be useful to express this relationship in the temporal frequency domain, because after taking Fourier transforms in time, the spatiotemporal problem collapses to a spatial problem at each temporal frequency. To do so, we take the temporal Fourier transform of both sides of (6.1) to obtain

$$\sigma_c \nabla^2 \tilde{u} = \tilde{i}_c^Z - \tilde{i}_c^s, \tag{6.2}$$

where the tilde denotes the temporal Fourier transform. By the definition of membrane impedance, the membrane voltage at each point is related to the current across the cone membrane impedance by $\tilde{u} = \tilde{i}_c^Z\, Z_c(f)$. This is a generalization of Ohm's law for sinusoidally varying current and voltage. With this relation, (6.2) can be rewritten as

$$\sigma_c \nabla^2 \tilde{u} = \frac{\tilde{u}}{Z_c} - \tilde{i}_c^s, \tag{6.3}$$

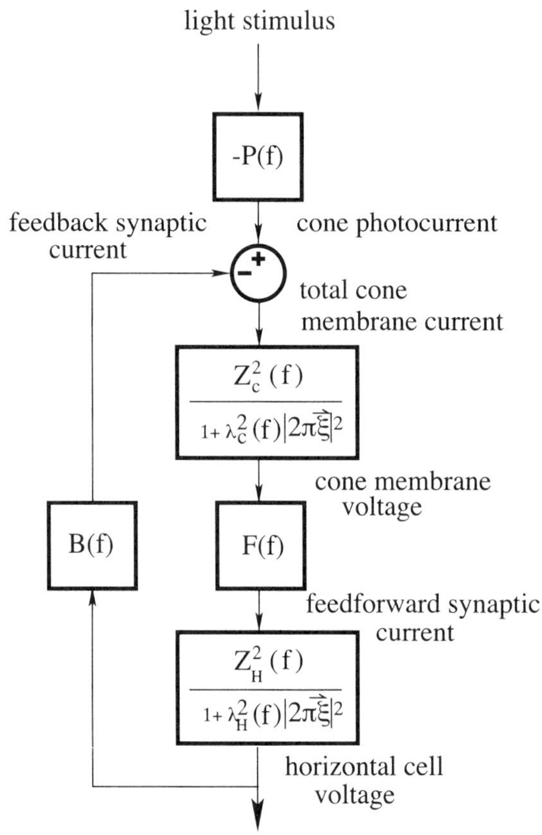

FIGURE 11. Circuit diagram of the cone-horizontal cell network.

Finally, we take the spatial Fourier transform of (6.3) and use the fact that the Fourier transform of a δ-function is 1, and the fact that the Fourier transform of $\nabla^2 \tilde{u}$ is $-|2\pi\vec{\xi}|^2 \hat{u}$, where the hat denotes the spatiotemporal Fourier transform. Thus

$$-\sigma_c \, |2\pi\vec{\xi}|^2 \hat{u} = \frac{\hat{u}}{Z_c} - \tilde{i}_c^s. \tag{6.4}$$

Multiplying (6.4) through by Z_c and putting all terms involving $\hat{u}$ on the same side of the equation gives

$$\left[1 + \lambda_c^2 \, |2\pi\vec{\xi}|^2\right] \hat{u} = Z_c \tilde{i}_c^s, \tag{6.5}$$

where $\lambda_c^2 = \sigma_c Z_c$ has units of length squared.

The relationship between source current, i_h^s, and membrane voltage, v, in the horizontal cell layer is of the same form:

$$\left[1 + \lambda_h^2 \, |2\pi\vec{\xi}|^2\right] \hat{v} = Z_h \, \tilde{i}_h^s, \tag{6.6}$$

where Z_h is the impedance per unit area of the horizontal cell membrane; $\lambda_h^2 = \sigma_h Z_h$, and σ_h is the conductance for current spread in the horizontal cell network.

The spatiotemporal transfer function of the cone and horizontal cell networks can now be found by putting into equations (6.5) and (6.6) explicit terms for the source currents. Let us denote the transfer function between the local light stimulus, $s(\vec{x},t)$, and the light-sensitive current, $i_L(\vec{x},t)$ by $-P(f)$; i.e. $\tilde{i}_L(\vec{x},f) = -P(f)\,\tilde{s}(\vec{x},f)$, and $\hat{i}_L(\vec{\xi},f) = -P(f)\,\hat{s}(\vec{\xi},f)$. Let us denote the transfer function between the local horizontal cell membrane potential, $v(\vec{x},t)$, and the feedback synaptic current, $i_B(\vec{x},t)$ "injected" into the cone network by $B(f)$; i.e. $\tilde{i}_B(\vec{x},f) = B(f)\,\tilde{v}(\vec{x},f)$, and $\hat{i}_B(\vec{\xi},f) = B(f)\,\hat{v}(\vec{\xi},f)$. The total source current in the cone network is now given by

$$i_c^s = i_L + i_B. \tag{6.7}$$

From the above definitions, it follows that

$$\hat{i}_c^s = -P\,\hat{s} + B\,\hat{v} \tag{6.8}$$

The source current for the horizontal cell network is of one type only, synaptic current from feed-forward synapses made by cones onto horizontal cells. If we denote the transfer function between cone membrane potential, $u(\vec{x},t)$ and synaptic current, $i_F(\vec{x},t)$, by $F(f)$, then $\hat{i}_F(\vec{\xi},f) = F(f)\,\hat{u}(\vec{\xi},f)$. When we identify the source current, i_h^s with i_F, we obtain

$$\hat{i}_h^s = F\,\hat{u} \tag{6.9}$$

Now we can put together equations (6.5), (6.6), (6.8) and (6.9), and solve for the Fourier transform of the cone voltage to obtain

$$\hat{u} = \left[\left(\frac{PZ_c}{1+\lambda_c^2|2\pi\vec{\xi}|^2}\right)\left(\frac{1}{1+\frac{FZ_hBZ_c}{(1+\lambda_c^2|2\pi\vec{\xi}|^2)(1+\lambda_c^2|2\pi\vec{\xi}|^2)}}\right)\right]\hat{s} \tag{6.10}$$

The term enclosed in square brackets multiplying $\hat{s}$ is the spatiotemporal transfer function of the cone; let us call it $K_{cone}(\vec{\xi},f)$:

$$K_{cone}(\vec{\xi},f) = \left(\frac{PZ_c}{1+\lambda_c^2|2\pi\vec{\xi}|^2}\right)\left(\frac{1}{1+\frac{FZ_hBZ_c}{(1+\lambda_c^2|2\pi\vec{\xi}|^2)(1+\lambda_c^2|2\pi\vec{\xi}|^2)}}\right) \tag{6.11}$$

Note that $K_{cone}(\vec{\xi},f)$ is function of $|\vec{\xi}|^2$; this reflects the radial symmetry of the cone's spatiotemporal impulse response function. A plot of $K_{cone}(\vec{\xi},f)$ at $f=0$ is shown on Figure 12.

Note that the cone responds best to intermediate spatial frequencies. This is a reflection in the frequency domain of the antagonistic center-surround receptive field organization of the cone. This organization of the cone receptive field is a consequence of the negative feedback onto cones from horizontal cells, which integrate input over large spatial areas. The antagonistic center-surround organization is also reflected in the *line-spread function* shown in Figure 13. The line-spread function gives the amplitude of the response to sinusoidal modulation of the contrast of a narrow bar as a function of the bar position. The line-spread function is obtained by evaluating the spatiotemporal transfer function at zero spatial frequency along one axis and then taking the inverse Fourier transform with respect to the other spatial frequency component. The line-spread function in Figure 13 is at zero temporal frequency. The corresponding stimulus is a steady bar, and the line-spread

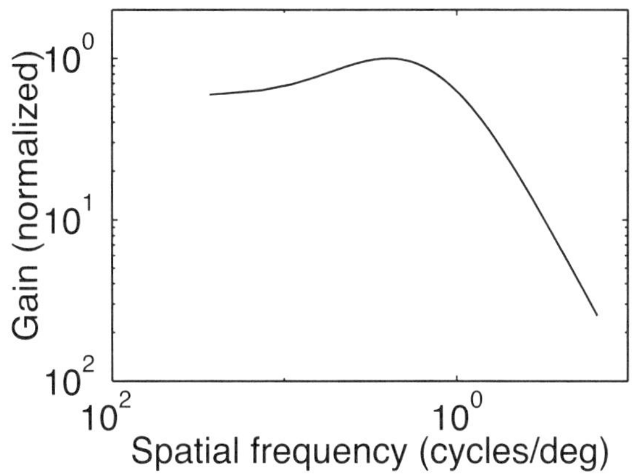

FIGURE 12. The cone spatiotemporal transfer function. The modulus (gain) of the cone spatiotemporal transfer function $K_{cone}(\vec{\xi}, f)$ (equation 6.10) , evaluated at temporal frequency, f equal to zero, plotted as a function of the magnitude of spatial frequency, $|\vec{\xi}|$.

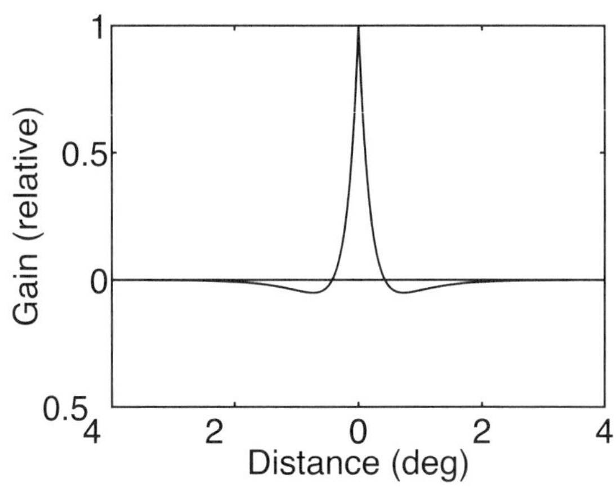

FIGURE 13. The cone line-spread function function. The line-spread (line-weighting) function, evaluated at temporal frequency, f equal to zero. This function gives the amplitude of the cone response to a narrow bar stimulus as a function of bar position.

function gives the steady increment/decrement in cone membrane potential as a function of bar position.

7. Bipolar Cell Receptive Fields

Bipolar cells receive input from numerous cones. If the bipolar cell is modeled as an isopotential compartment, then the bipolar cell simply integrates all of the synaptic input current from cones and produces a voltage response determined by

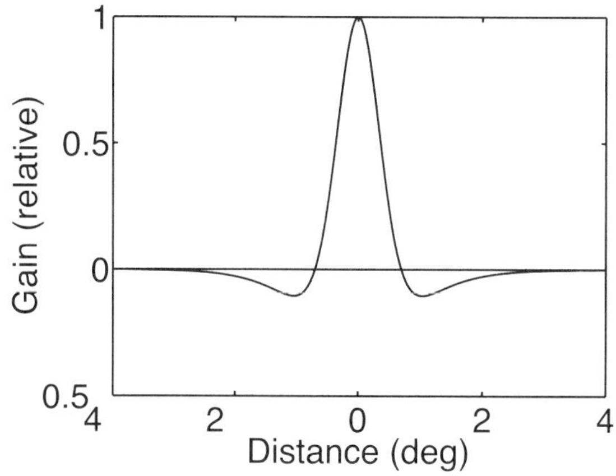

FIGURE 14. The bipolar cell line-spread function. The bipolar cell line-spread function function, evaluated at temporal frequency, f equal to zero. This function gives the amplitude of the bipolar cell response to a narrow bar stimulus as a function of bar position. Note the difference-of-Gaussians-like shape.

its equivalent membrane impedance, $Z_b(f)$. In general, the strength of synaptic input to a neuron falls off with distance between that neuron and its presynaptic neurons. This is a consequence of branching of the dendritic tree and increasing sparseness of the branches with distance from the cell body. If we assume that the synaptic weight of cone input to the bipolar cell falls off as a Gaussian function of distance from the receptive-field center with variance σ_b^2 , then the spatiotemporal transfer function of the bipolar cell, $K_{bipolar}(\vec{\xi}, f)$ is of the form

$$K_{bipolar}(\vec{\xi}, f) = Z_b(f) \exp\left(\frac{-\sigma_b^2|\vec{\xi}|^2}{2}\right) K_{cone}(\vec{\xi}, f) \tag{7.1}$$

A plot of the bipolar cell line-spread function corresponding to equation (7.1) is shown in Figure 14. The retinal ganglion cells receive synaptic input from several bipolar cells, and thus inherit the basic receptive field structure shown in Figure 14. The result is a difference-of-Gaussians receptive field structure like that shown in Figure 7.

8. Mechanisms of Adaptation: Future Challenges

The retina as a whole is a highly nonlinear system, in large part as a consequence of various mechanisms of adaptation. One form of adaptation is referred to as *light adaptation*. Mechanisms of light adaptation act to adjust the sensitivity, kinetics and spatial aspects of the response properties of visual neurons with changes in the level of ambient illumination. There are a number of stages of light adaptation in the retina. A great deal of progress has been made recently in understanding mechanisms of light adaptation in the phototransduction process in rods and cones. Mathematics has played and continues to play a very important role in the development and testing of models for phototransduction [**7, 15**]. These models are based on the underlying biochemistry and membrane biophysics. The other stages of light adaptation in the retina, sometimes referred to as network adaptation, are not understood nearly as well.

Another form of adaptation is referred to as *contrast gain control* [**13**]. In this form of adaptation, response kinetics are adjusted according to the recent history of contrast at a given fixed mean light level. It appears likely that amacrine cells play an important role in this form of adaptation. Both forms of adaptation involve nonlinear feedback.

Many of the remaining challenges in retinal physiology and in visual physiology in general involve the characterization and modeling of neurons with nonlinear response properties. There are great opportunities for physiologists and mathematicians to work together in an effort to elucidate the neural mechanism underlying visual perception and to develop accurate mathematical models.

References

[1] S. E. Brodie, B. W. Knight and F. Ratliff, *The response of the Limulus retina to moving stimuli: a prediction by Fourier synthesis*, Journal of General Physiology, **72** (1978), pp. 129–165.

[2] H. K. Hartline and F. Ratliff, *Spatial summation and inhibitory influences in the eye of Limulus, and mutual interaction of receptor units*, Journal of General Physiology, **41** (1958), pp. 1049–1066.

[3] A. L. Hodgkin and A. F. Huxley, *A quantitative description of membrane current and its application to conduction and excitation in nerve*, Journal of Physiology, **117** (1952), pp. 500–544.

[4] D. A. Kraaij, H. Spekreijse and M. Kamermans, *The nature of surround-induced depolarizing response in goldfish cones*, Journal of General Physiology, **115** (2000), pp. 3-15.

[5] H. I. Krausz and K.-I. Naka, *Spatiotemporal testing and modeling of catfish retinal neurons*, Biophysical Journal, **29** (1980), pp. 13–36.

[6] S. W. Kuffler, *Discharge patterns and functional organization of mammalian retina, Journal of Neurophysiology*, **16** (1953), pp. 37–68.

[7] R.D. Hamer, *Computational analysis of vertebrate phototransduction: combined quantitative and qualitative modeling of dark- and light-adapted responses in amphibian rods*, Visual Neuroscience, **17** (2000), pp. 679–99.

[8] C. S. Peskin, D. Tranchina and D. Hull, *How to see in the dark: Photon noise in vision and nuclear medicine*, Annals of the New York Academy of Sciences, **435** (1984), pp. 48–72.

[9] F. Ratliff, *Mach Bands: Quantitative studies on neural networks in the retina*, Holden Day, San Francisco, CA, 1965.

[10] R. W. Rodieck, *Quantitative analysis of cat retinal ganglion cell responses to visual stimuli*, Vision Research, **5** (1965), pp. 583–601.

[11] R. W. Rodieck, *The First Steps in Seeing*, Sinauer Associates, Sunderland, MA, 1998.

[12] M. Schetzen, *The Volterra and Wiener Theories of Nonlinear Systems*, Robert E. Krieger Publishing Company, Malabar, FL, 1989.

[13] R.M. Shapley and J.D. Victor, *How the contrast gain control modifies the frequency responses of cat retinal ganglion cells*, Journal of Physiology, **318**, (1981), pp. 161–79.

[14] D. Tranchina, J. Gordon and r. Shapley, *Spatial and temporal properties of luminosity horizontal cells in the turtle retina*, Journal of General Physiology, **82**, (1983), pp. 573–598.

[15] D. Tranchina, J. Sneyd J, and I. D.Cadenas, *Light adaptation in turtle cones. Testing and analysis of a model for phototransduction*, Biophysical Journal, **60**, (1991), pp. 217–37.

[16] J. D.Victor, *Nonlinear systems analysis in vision: overview of kernel methods* In *Nonlinear Vision: Determination of Neural Receptive Fields, Function, and Networks*, ed. R. Pinter and B. Nabet. CRC Press, Cleveland OH, 1992, pp. 1-37.

[17] B. A. Wandell, *Foundations of Vision*, Sinauer Associates, Sunderland, MA, 1995.

Department of Biology, New York University, New York, NY 10003, Courant Institute of Mathematical Sciences, 251 Mercer St., New York, NY 10012

E-mail address: tranchina@cims.nyu.edu

Proceedings of Symposia in Applied Mathematics
Volume **59**, 2002

Arrhythmias by Dimension

James P. Keener

ABSTRACT. The mathematical study of cardiac arrhythmias can be organized by classifying arrhythmias by their spatial dimension. This paper gives an overview of cardiac arrhythmias from this organizational viewpoint, emphasizing the insights that are gained and the problems that result from identifying the arrhythmias in these terms.

1. Introduction

Abnormalities of function of the cardiac conduction system are the cause of death of thousands of people every day. For that reason, the study of cardiac arrhythmias is of great interest from a medical and scientific perspective. However, cardiac arrhythmias are also interesting for mathematical reasons because they can be studied from the viewpoint of dynamical systems theory.

Strictly speaking, a cardiac arrhythmia is any departure of the heartbeat from strict periodicity. In that sense, all living persons have arrhythmias all the time, since the normal heartbeat is never exactly periodic. However, the arrhythmias of interest here are those which are considered abnormal, and which have nontrivial physiological consequences.

There are (at least) three challenges presented by cardiac arrhythmias. The first is simply to identify the different arrhythmias and their underlying mechanisms. The second is to understand the cause or origin of each arrhythmia, and the third is to determine how to control the arrhythmia so that its harmful effects can be avoided. To the mathematician, there is also the challenge and fascination with the mathematics that underlies the description and understanding of arrhythmias.

The purpose of this paper is to provide a brief introduction to each of these, organized on the basis of their spatial dimension.

2. The Normal Cardiac Rhythm

Before we can discuss cardiac arrhythmias, we must have a description of the normal course of events.

1991 *Mathematics Subject Classification.* 92C30, 92C05, 92C50, 35K57.
Key words and phrases. cardiac arrhythmias, reentry, spiral waves, scroll waves.
This work was supported in part by NSF Grant DMS-99700876.

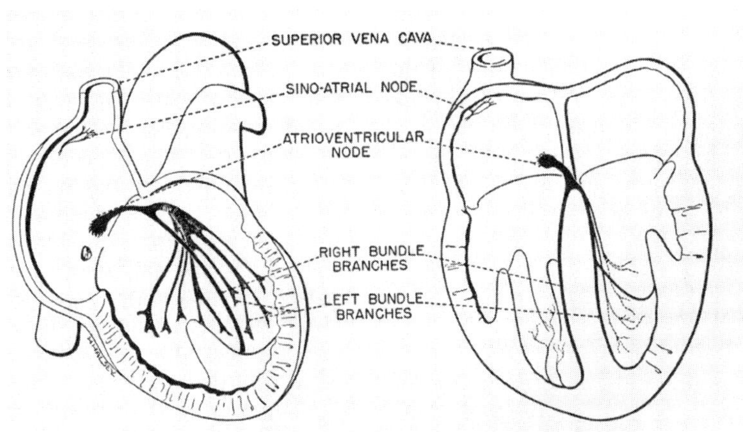

FIGURE 1. Schematic diagram of the cardiac conduction system. ([**Ru76**], Fig. 3-9, p. 87)

The human heartbeat is an electrical event, occurring about once per second, during which every cell in the heart experiences a rapid excursion of transmembrane potential , lasting about 300 ms, before returning to rest. This excursion from rest, called an action potential , is coordinated by a wave of electrical activity. This wave of electrical activity (see Fig. 1) is initiated by a small collection of cells in the right atrium, near the superior vena cava, called the sinoatrial node (SA) node. These cells are all autonomous oscillators, and are synchronized by local electrical coupling. Once initiated, a wave of electrical activity propagates across the atria, terminating at the atrial-ventricular (AV) septum, a non-conducting septum between atria and ventricles. This wave of activity enters another small clump of cells in the AV septum, the AV node, through which it propagates slowly. Leaving the AV node, the wave enters a bundle of specialized fibers called Purkinje fibers (also called the bundle of HIS), which branch out in tree-like fashion to reach the interior (endocardial) wall of the ventricles. The excitation emerges from the Purkinje fibers to activate the ventricular tissue, propagating from inside to outside, terminating at the epicardium.

The standard diagnostic tool for the clinical evaluation of the heartbeat is the electrocardiogram (ECG) , which measures the body surface potential generated by the traveling heartbeat (see Fig. 2). The normal ECG has three characteristic signals, the P wave, corresponding to activation of the atria, the QRS complex, corresponding to the activation of the ventricles, and the T wave, corresponding to recovery (return to rest) of the ventricles. Other electrical events involve too little tissue mass to be recorded at the body surface.

3. Zero Dimensional Arrhythmias

Zero dimensional arrhythmias are those involving single cells or small collections of cells, and do not rely on the spatial arrangement of the cells to exist. These can be classified roughly into three categories. The first are single cells that change their behavior from normal behavior to a different type of rhythmic or arrhythmic behavior. For example, SA nodal cells are normally autonomous oscillators while

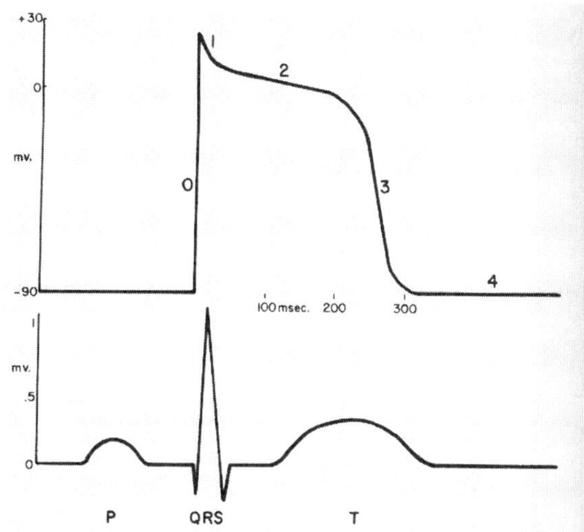

FIGURE 2. Cellular transmembrane potential and electrocardiogram. The upper tracing represents the transmembrane potential of a single ventricular myocyte and the lower tracing shows the body surface potential during the same electrical event. The numbers on the upper tracing designate phases in the action potential cycle: 0: the upstroke, 1: the brief spike, 2: the plateau, 3: the rapid recovery, 4: resting potential. ([**Ru76**], Fig. 8-4, p. 286)

most other myocardial cells do not spontaneously fire without external stimulus. An ectopic focus is a collection of cells that spontaneously oscillates although it should not. Similarly, some cells may experience a change from their normal action potential trajectory. For example, early after depolarizations (EAD's) occur when cells do not return directly to rest after having been stimulated, but spontaneously fire before returning to rest [**LR94b**].

The second class of zero dimensional arrhythmias are those associated with coupled oscillators. For example, parasystole is a condition in which there are (purportedly) two autonomous oscillators vying for control of the heartbeat.

The third class of zero dimensional arrhythmias are those associated with failure of a group of cells to be entrained to an external signal. AV nodal block is usually viewed in this way.

The mathematical study of these arrhythmias makes use of dynamical systems theory to characterize the possible behaviors and bifurcations. There are two primary approaches. In the first, detailed (quantitative) ionic models of single cells are studied to determine how physical parameters affect the behavior of a cell in a variety of conditions. There is a long history of modeling studies of this type. The first detailed model of cardiac cell activity was proposed by Noble [**Nob62**], and was a modification of the Hodgkin-Huxley model [**HH52**]. This style of modeling was continued by many authors, including [**DN85**] for Purkinje fiber, [**BR77**] for ventricular cells, and [**YNI80**] for SA nodal cells. As more information has become available concerning the complexities of cardiac cells, the models have become correspondingly more complex, as represented by [**LR94a**] and [**JRW98**] for ventricular cells, [**WKVN93**] for SA nodal cells, [**CRN98**] for atrial cells.

Much of the current research activity is aimed at understanding cellular behavior in a pathological environment, such as ischemia, which occurs immediately following a heart attack. There are a large number of changes that take place [**Car99**], and we are a long way from having a clear mathematical description of these events. Detailed models have been used to study a variety of related abnormalities. For example, [**LR94b**] examines EAD's, [**WKV+93**] examines ectopic rhythms, [**CVJCN98**] examines pH regulation in ischemic cells, [**FSFT96**] examines the role of K-ATP channels during ischemia, and [**RJW99**] examines irregularities in the dynamics of calcium handling.

3.1. The Poincaré Oscillator. There are numerous challenges associated with the qualitative behavior of a zero dimensional arrhythmia. The most obvious is simply to describe the different possible behaviors of an oscillator and the bifurcations between these behaviors.

The study of cellular arrhythmias from a qualitative perspective has a long history, initiated by the famous van der Pol equation, initially proposed as a model of the heartbeat [**vdPvdM28**]. In more recent times there have been many studies of coupled oscillators and forced oscillators that take into account a range of features relevant to cardiac cell activity.

A simple model of forced periodic oscillations, called the Poincaré oscillator (also called a radial isochron clock or snap back oscillator) manages to portray many of the important ideas. We suppose that a point is moving counterclockwise around a circle of radius 1 at fixed velocity $\frac{d\theta}{dt} = 1$. At some point in its rotation, the point is moved horizontally by an amount A and then allowed to instantly "snap back" to radius 1, moving along a radial line toward the origin, without additional change of phase.

It is fairly easy to determine the relationship between the phase ϕ immediately after reset as a function of phase θ immediately before reset. This relationship is called the phase resetting curve for this process, and depending on the amplitude of A can be of two distinct topological types. If $A < 1$, then $\phi(\theta + 2\pi) = \phi(\theta) + 2\pi$, and is called a type 1 map. If $A > 1$, then $\phi(\theta + 2\pi) = \phi(\theta)$, and is called a type 0 map.

Having determined the phase resetting curve, one can now study the long term behavior of this oscillator by examining the map

$$\theta_{n+1} = \phi(\theta_n) + T. \tag{3.1}$$

It is a straightforward matter to find the parameter values for which there is one-to-one phase locking, but there are many more complicated response patterns that are much more difficult to unravel. In fact sorting out the details of this apparently simple map has been the focus of numerous papers, including [**GM79**] [**GP82**] [**GG82**] [**GGSP83**] [**KG84**] [**GGBS84**] [**MFV88**] [**GS94**].

One very important idea that has emerged from these studies is the idea of phase resetting and phase resetting curves, allowing cardiac rhythms to be studied as iterates of maps [**LGBP92**] [**GG99**]. This idea was important, because phase resetting curves can be measured experimentally and their consequences examined

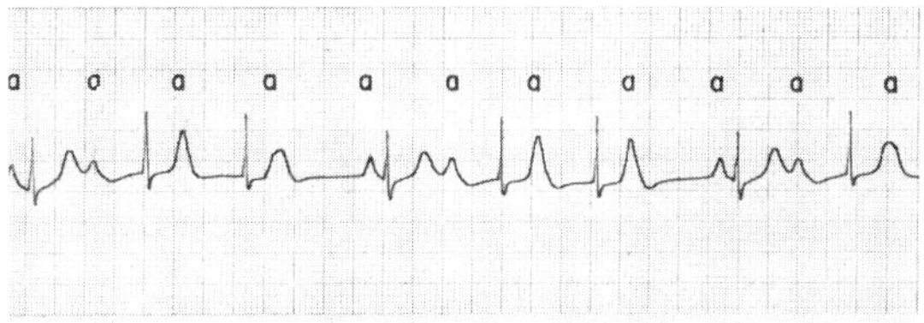

FIGURE 3. ECG recording of a Wenckebach pattern in which every fourth or fifth atrial beat is not conducted. The symbol "a" denotes an atrial excitation (P-wave).([**Ru76**], Fig. 8-24, p. 313)

without reference to a detailed ionic model [**JA79**]. In this way, situations that are far more complicated than the simple Poincaré map can be studied and analyzed.

3.2. AV node. As an example of how dynamical systems theory can be used to understand specific cardiac arrhythmias, we consider AV nodal block. In the normal heart, the only pathway for an action potential to travel to the ventricles is through the AV node. As noted above, propagation through the AV node is slow compared to propagation in other cardiac cells, and as a result is somewhat more likely to fail.

Propagation failure in the AV node leads to skipped QRS complexes on the ECG, or, more prosaically, skipped heartbeats. A skipped heartbeat once in a while is not particularly dangerous, but it is certainly noticeable. During diastole (the period of ventricular relaxation during the heartbeat cycle), the ventricles fill with blood. Following an abnormally long diastolic period, the heart becomes enlarged and when the next compression (systole) occurs, Starling's law (i.e., that compression is stronger when the heart is more distended initially) takes control and compression is noticeably more vigorous, giving the subject a solid thump in the chest.

AV nodal conduction abnormalities are classified into three classes that are all readily visible from ECG recordings by looking at the time interval between the P wave and the QRS complex, i.e., the P-R interval. Type I AV nodal block shows itself as an increase in the P-R interval as the SA pacing rate increases. Type III AV nodal block corresponds to no AV nodal conduction whatever, and total absence of a QRS complex.

Type II AV nodal block is phenomenologically the most interesting. In the simplest type, there is one QRS complex for every two P waves, a 2:1 pattern. A more complicated pattern is as follows: On the ECG (Fig. 3) P waves remain periodic, although the P-R interval is observed to increase gradually until one QRS complex is skipped. Following the skipped beat, the next P-R interval is quite short, but then the P-R lengthening begins again leading to another skipped beat, and so on. A pattern with n P waves to $n-1$ QRS complexes is called an n to $n-1$ Wenckebach rhythm, after the German cardiologist Wenckebach [**Wen04**] .

A simple mathematical description of AV nodal signal processing can be given as follows: we view the AV node as a collection of cells that fire when they are excited, which happens if their potential reaches a threshold, $\theta(t)$. Immediately after firing, the cells become refractory, but then gradually recover. Effectively, at firing, the threshold increases dramatically, but then decreases back to its steady state value as recovery proceeds. This model ignores the fact that the AV node is self-oscillatory, and will fire without stimulus with a low frequency of 30-40 per minute. The self-oscillatory nature of the AV node becomes evident only in cases of SA nodal failure or at very low SA nodal firing rates. Thus, the model discussed here is valid at high stimulus rates (appropriate for AV nodal block), but not at low stimulus rates.

Input to the AV node comes from the action potential propagating through the atria from the SA node. The AV node experiences a periodic, time varying potential, say $\phi(t)$. Firing occurs if the input signal reaches the threshold. Therefore at the nth firing time, denoted t_n,

$$\phi(t_n) = \theta(t_n). \tag{3.2}$$

Subsequent to firing, the threshold evolves according to

$$\theta(t) = \theta_0 + [\theta(t_n^+) - \theta_0]e^{-\gamma(t-t_n)}, \qquad t > t_n. \tag{3.3}$$

Note that $\theta \to \theta_0$ as $t \to \infty$, and thus θ_0 denotes the base value of the threshold. Further, $\theta = \theta(t_n^+)$ at $t = t_n$, and thus $\theta(t_n^+) - \theta(t_n^-)$ denotes the jump in the threshold caused by the firing of an action potential. To complete the model we must specify $\theta(t_n^+)$. The important feature of $\theta(t_n^+)$ is that it must have some memory, that is, depend in some way on $\theta(t_n^-)$. Therefore, we take

$$\theta(t_n^+) = \theta(t_n^-) + \Delta\theta. \tag{3.4}$$

The simplest choice is to take $\Delta\theta$ a constant.

Now we can find the next firing time as the smallest solution of the transcendental equation

$$\phi(t_{n+1}) = \theta_0 + [\theta(t_n^+) - \theta_0]e^{-\gamma(t_{n+1}-t_n)}. \tag{3.5}$$

Equation (3.5) can be rearranged into an equation of the form

$$F(t_{n+1}) = F(t_n) + \Delta\theta e^{\gamma t_n} = G(t_n), \tag{3.6}$$

where

$$F(t) = (\phi(t) - \theta_0)e^{\gamma t}. \tag{3.7}$$

Plots of typical functions $F(t)$ and $G(t)$ are shown in Fig. 4. Here we have taken $\phi(t) - \theta_0 = \sin^4(\pi t)$. The dashed lines in this figure follow a few iterates of the map.

The key observation is that the map $t_n \mapsto t_{n+1}$ as defined by (3.6) is the lift of a . A circle map is a map of the circle to itself, $f : S^1 \to S^1$, but it is often easier to describe a circle map in terms of its lift $F : R \to R$ where F is a monotone increasing function, and $F(x + 1) = F(x) + 1$. The two functions f and F are related by

$$f(x) = F(x \bmod 1) \bmod 1. \tag{3.8}$$

(For convenience we normalize the circumference of the circle to be of length 1, rather than 2π.)

Notice that the map defined by (3.6) is indeed the lift of a circle map, since, if t_n and t_{n+1} satisfy (3.6), then so do $t_n + T$ and $t_{n+1} + T$. To find a circle map, we let k_n be the largest integer less than t_n/T and define $\psi_n = (t_n - k_n T)/T$. In these variables the map (3.6) can be written as

$$f(\psi_{n+1}) = (f(\psi_n) + \Delta\theta e^{\gamma T\psi_n})e^{\gamma T\Delta k_n}, \tag{3.9}$$

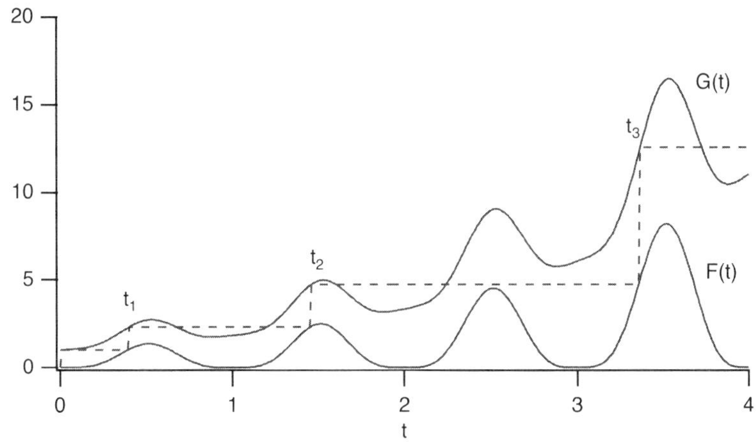

FIGURE 4. Plot of the functions $F(t)$ and $G(t)$ with $\Delta\theta = 1.0, \gamma = 0.6$.

where

$$f(\psi) = (\Phi(\psi) - \theta_0)e^{\gamma T\psi}, \qquad \Phi(\psi) = \phi(T\psi), \tag{3.10}$$

and $\Delta k_n = k_{n+1} - k_n$.

We can make a few observations about the map $\psi_n \mapsto \psi_{n+1}$. First, and most disconcerting, the map is not continuous. In fact, it is apparent that there are values of t on the unit interval that can never be firing times. For t to be permitted as a firing time it must be the first point at which $F(t)$ reaches the level $G(t_n)$, i.e., the first time that the threshold is reached. At such a point, $F'(t) > 0$. Since there are regions for which $F'(t) < 0$ which can therefore never be firing times, this is a map of the unit interval *into*, but not onto, itself. However, the map $t_n \mapsto t_{n+1}$ is order preserving since $G(t)$ is increasing whenever $F(t)$ is increasing.

Since the entire unit interval is not covered by the map, it is only necessary to examine the map on its range. Examples of the map $\psi_n \mapsto \psi_{n+1}$ are shown in Figs. 5–8. Here we have plotted the map only on the attracting range of the unit interval. These show important and typical features, namely that the map consists of either one or two continuous, monotone increasing branches. The first branch, with values above the one-to-one curve, corresponds to firing in response to the subsequent input (with $k_{n+1} = k_n + 1$), and the second, with values below the one-to-one curve, corresponds to firing after skipping one beat (with $k_{n+1} = k_n + 2$). The skipped beat occurs because when the stimulating pulse arrives, it is subthreshold and so does not evoke a response.

The sequence of figures in Figs. 5–8 is arranged in order of decreasing values of γT. Note that as γ decreases, the rate of recovery from inhibition decreases. For γT sufficiently large, there is a unique fixed point, corresponding to firing in 1:1 response to the input signal. This makes intuitive sense, for when γ is large the recovery from inhibition is fast, and thus the AV node can be driven at the frequency of the SA node. For large γT the map is relatively insensitive to changes in parameters. As γT decreases the first branch of the map increases and the value of the fixed point increases, corresponding to a somewhat delayed firing. Furthermore, because the slope of the map in the vicinity of the fixed point is close

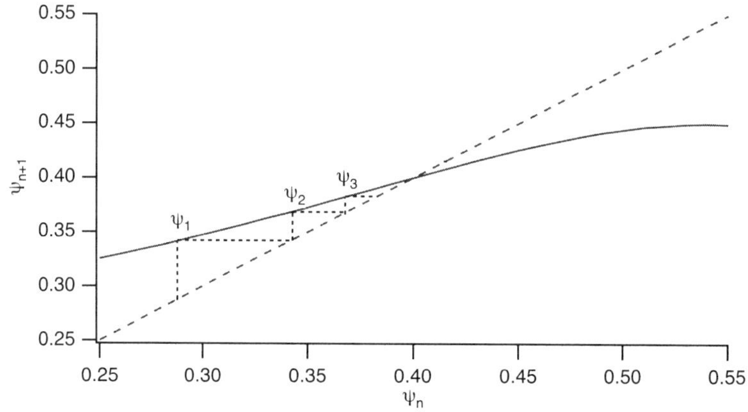

FIGURE 5. Plot of the map $\psi_n \mapsto \psi_{n+1}$ with $\Delta\theta = 1.0$, with $\gamma T = 0.8$.

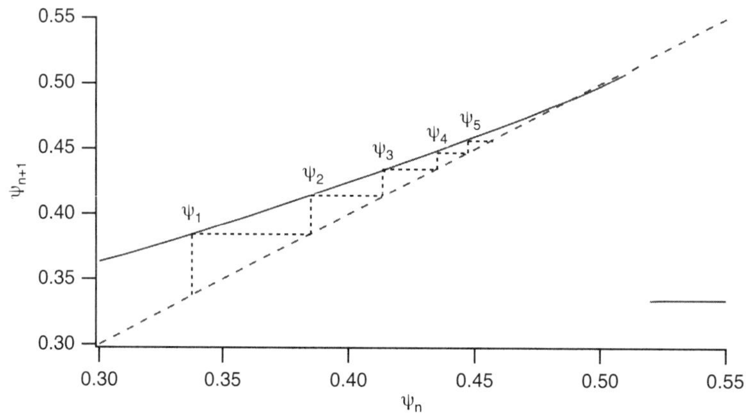

FIGURE 6. Plot of the map $\psi_n \mapsto \psi_{n+1}$ with $\Delta\theta = 1.0$, with $\gamma T = 0.695$.

to 1, the fixed point is sensitive to changes in parameter values (depicted in Fig. 6), corresponding to type I AV block.

As the parameter γT is decreased further, the fixed point is lost and a second branch to the map appears (as in Fig. 7). Iterations show that subsequent firings become later and later in the input cycle until one beat is skipped, followed by a firing which is relatively early in the input cycle. For this region of parameter space, the map replicates the Wenckebach phenomenon.

Finally, as γT decreases further the second branch "slides over" to the left and eventually intersects the one-to-one line, yielding a fixed point. This fixed point corresponds to a periodic pattern of one skipped beat for each successful firing, a two-to-one pattern, and replicates type II AV block.

The behavior of the map in the region with no fixed point can be described by the rotation number . For maps of the type (3.6) the rotation number can be

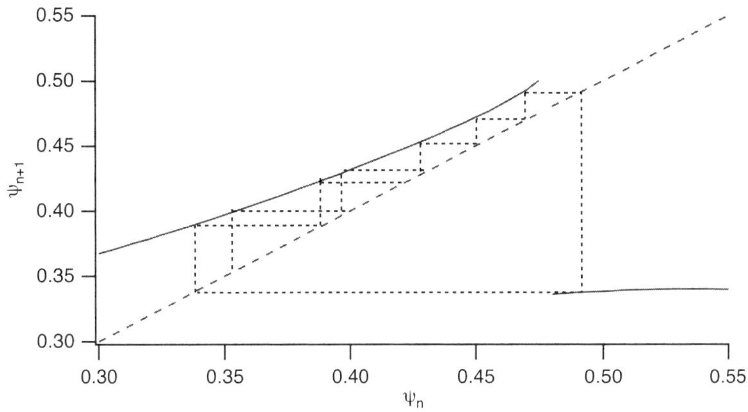

FIGURE 7. Plot of the map $\psi_n \mapsto \psi_{n+1}$ with $\Delta\theta = 1.0$, with $\gamma T = 0.67$.

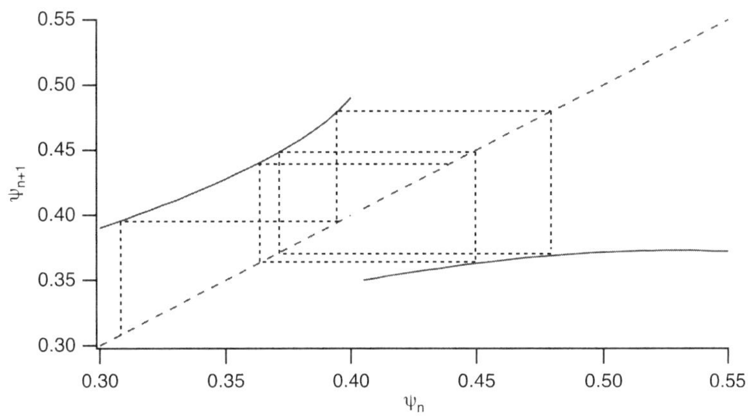

FIGURE 8. Plot of the map $\psi_n \mapsto \psi_{n+1}$ with $\Delta\theta = 1.0$, with $\gamma T = 0.55$.

defined by

$$\rho = \lim_{n \to \infty} \frac{t_n}{nT}. \tag{3.11}$$

A detailed exposition of the rotation number for continuous circle maps can be found in chapter 17 of [**CL84**]. For circle maps that are not continuous, the following features of the rotation number ρ can be verified [**Kee81**] [**Kee80**].

1. ρ exists and is independent of initial data.
2. ρ is a monotone decreasing function of γT.
3. ρ attains *every* rational level between 0 and 1 on an open interval of parameter space.

For continuous circle maps, it is not certain that every rational level is attained on an open interval of parameter space.

The main consequence of this result is that between 1:1 phase locking and 2:1 AV block, for every rational number there is an open interval of γT on which the rotation with that rational number is attained. The last two features of ρ make the behavior of the orbits so unusual, being a function that is monotone non-decreasing, yet locally constant at all the rational levels. Such a function is called the Devil's Staircase. Notice that if ρ is rational on an open interval of parameter space, then phase locking is robust.

This model of AV nodal conduction block is quite simple. Generalizations of it, using clinical data, have found maps that are nonmonotonic circle maps having correspondingly more complicated phenomena [**SAGB95**] [**BAZS94**]. Other examples of how circle maps can be applied to biological rhythms include [**Kni72**], [**GM79**] [**KHR81**].

3.3. Pacing and Alternans. The ideas of iterates of maps can also be used to study the response of single atrial or ventricular cells to periodic stimuli. We suppose that a cell is stimulated by a brief pulse at a fixed period, called the basic cycle length (BCL). This basic cycle length is divided between two types of cellular activity, its action potential and its recovery. Clearly the action potential duration (APD) during the nth cycle, APD_n, and following recovery time, RT_n, must sum to the basic cycle length

$$APD_n + RT_n = BCL. \tag{3.12}$$

The fundamental assumption (an approximation, at best) is that APD_n is a function of the *previous* recovery time RT_{n-1}, $APD_n = a(RT_{n-1})$, called the restitution curve . It follows that

$$RT_n = BCL - a(RT_{n-1}), \tag{3.13}$$

provided there are no skipped beats. The restitution curve is typically monotone increasing [**FSS$^+$83**] [**WOG95**] [**YJAG99**] and the bifurcations of this map can be studied using standard methods. Two simple observations are, first, that if a is monotone increasing, then, if it exists, a fixed point is unique, and second, the fixed point is stable if and only if $a' < 1$ at the fixed point. Since $a' \to 0$ for large RT, the response to a periodic stimulus is aperiodic only if the BCL is small, i.e., the stimulus is rapid. Notice also that if the solution is unstable, then iterates alternate above and below the fixed point. These oscillations of APD are called alternans in the cardiology literature.

These methods have also been applied to models of other arrhythmias, specifically parasystole, as in [**MJMM77**], [**AM80**], [**GGB86**], [**JML86**], [**GGCS87**], [**CGRG89**], [**CGJ90**]. Additional discussions of maps and chaos and the like with application to a wide array of biological oscillators can be found in [**GM88**], [**GK95**], [**Str94**].

4. One-Dimensional Arrhythmias

One-dimensional arrhythmias are characterized as such because their existence relies in a crucial way on a one-dimensional path of propagation.

The classic example of a one-dimensional arrhythmia is associated with Wolff-Parkinson-White (WPW) Syndrome. With this syndrome, it is observed that the action potential enters the ventricles through two conducting pathways, the normal AV nodal pathway, and a second pathway through the AV septum, called an accessory pathway. If this is the case, the AV node could fail to conduct a pulse,

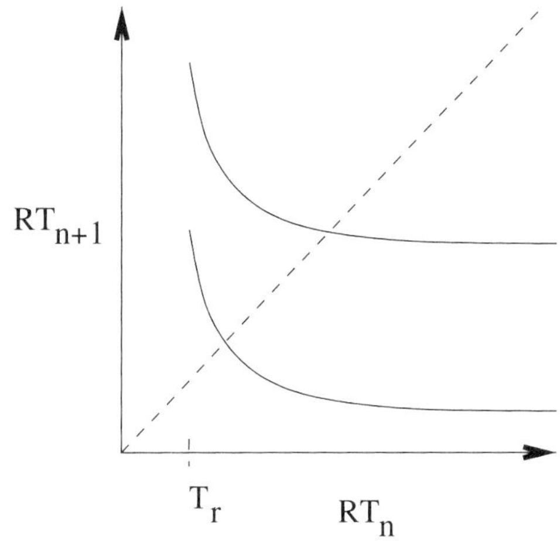

FIGURE 9. Schematic diagram of the map (3.13) for two different values of basic cycle length. For the larger BCL, the fixed point is stable, while for the smaller BCL, the fixed point is unstable, since $a' > 1$ at the fixed point.

while the activation could enter the ventricles through the accessory pathway, propagate retrogradely along the Purkinje fibers to the AV node thence into the atria to complete a circuit. A permanently circulating wavefront results in a very rapid heartbeat with inefficient pumping activity, and would likely lead to fainting or dizziness.

One-dimensional reentrant arrhythmias were first studied by Mines [**Min14**] when he intentionally cut a ring of tissue from around the superior vena cava and managed to initiate waves that traveled in only one direction.

One-dimensional arrhythmias are the easiest of all arrhythmias to treat clinically. If they can be identified as such, one-dimensional reentrant arrhythmias are treated by permanently burning a point on the pathway through which the circulating wave must go, thereby breaking the circuit. Thus, the important issues concerning one dimensional arrhythmias are to determine how they are formed, if they are stable once formed, and how to eliminate them. More important from a clinical point of view is to ascertain when an arrhythmia is truly one-dimensional, how to locate a narrow pathway through which the circuit passes, and how to destroy the circuit without damaging other important parts of the cardiac conduction system (such as the AV node).

We can get an idea of how a one-dimensional reentrant arrhythmia is initiated using a simple model. Suppose that there are cells located next to the exit from a one-dimensional path with one-way-block (Fig. 10). A path with one-way block means that propagation can go through the cells in one direction but not in the opposite direction. While this is not typical, it is certainly possible if there is some sort of arborization or branching of a pathway [**Pau82**] [**Kee84**] [**Kee87**] [**LG91**]. Suppose further that these cells are normally stimulated by some external pacemaker, with period T. We define the instantaneous frequency of stimulus as $\Delta T_{n+1} = t_{n+1} - t_n$, where t_n is the nth firing. Now we take a simple kinematic description of propagation in the one-way-path, and suppose that the speed of propagation in the path is a function of the instantaneous period, $c = c(\Delta T)$, called the dispersion curve. (Like the restitution curve, c is typically an increasing function of ΔT [**MR81**].) Then the travel time around the one-way loop is $\frac{L}{c(\Delta T)}$. In cardiac tissue, the speed of an action potential is on the order of 0.5 m/s, so that travel time around the loop is much shorter than the period of external stimulus. Thus the wave on the loop (typically) returns to the stimulus site long before the next external stimulus arrives (i.e. we assume that $L/c < T$). If this travel time is larger than the absolute refractory period T_r of the cells but smaller than T, the period of the external stimulus, then it stimulates the cells and initiates another wave around the loop. Thus,

$$\Delta T_{n+1} = t_{n+1} - t_n = \frac{L}{c(\Delta T_n)}, \tag{4.1}$$

provided $T > \frac{L}{c(\Delta T_n)} > T_r$.

On the other hand, if this travel time is smaller than T_r, the reentering stimulus is not successful, and the cells must await the next external stimulus before they fire, so that

$$\Delta T_{n+1} = t_{n+1} - t_n = T \tag{4.2}$$

if $\frac{L}{c(\Delta T_n)} < T_r$.

With this information, we can construct the map $\Delta T_n \to \Delta T_{n+1}$, (shown in Fig. 11). There are obviously two branches for this map (shown as solid curves). Of interest are the fixed points of this map, corresponding to a periodic pattern of stimulus. The fixed point on the upper branch corresponds to the normal stimulus pattern from the external source, whereas the fixed point on the lower branch corresponds to a high frequency, reentrant, pattern. The key feature of this map is that there is hysteresis between the two fixed points. In a "normal" situation (Fig. 11a), with L small and T large, the period of stimulus is fixed at T. However, as L increases or as T decreases, rendering $L > T_r c(\delta T_n)$, there is a "snap" onto the smaller period fixed point, corresponding to initiation of a reentrant pattern (Fig. 11b). The pernicious nature of the reentrant pattern is demonstrated by the fact that increasing the period of the external pacemaker back to previous levels does not restore the low frequency pattern - the iterates of the map stay fixed at the lower fixed point, even though there are two possible fixed points. This is because the circulating pattern acts as a retrograde source of high frequency stimulus on the original stimulus site, thereby masking its periodic activity. The reentrant oscillator obstructs the activity of the external pacemaker (much as in parasystole).

Note that there are a number of ways that this reentrant pattern might be initiated. First, following a heart attack, a growing infarcted region may lead to a gradual increase in L, initiating the reentrant pattern while keeping T fixed. On the other hand, an infarcted region may exist but remain static (L fixed), and the reentrant pattern is initiated following a decrease in T, for example, during strenuous exercise. Thus, a static one-way-loop acts like a "frequency bomb" (rather than a time bomb), ready to go off whenever the frequency of stimulus is sufficiently high.

The analysis of waves on a one-dimensional ring is substantially easier and more convenient than reentrant waves in higher dimensional domains. For that reason

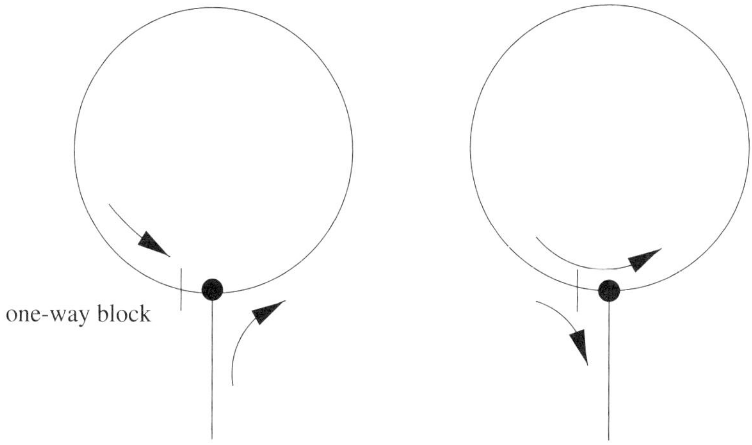

FIGURE 10. Diagram of a conducting path with one way block, preventing conduction from right to left. Left panel: Conduction of a stimulus around the loop until it encounters refractoriness and fails to propagate further. Right panel: Conduction of a reentrant pattern circulating continuously around the loop, and exiting via the entry pathway on every circuit.

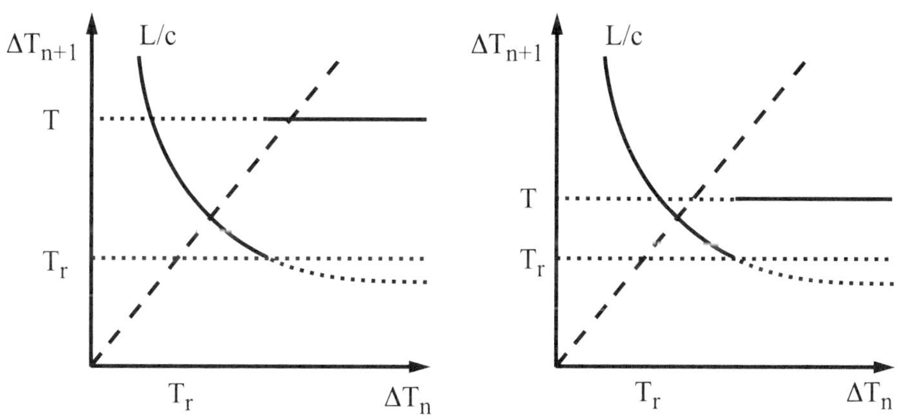

FIGURE 11. Next interval map for a one way conducting loop in two cases. Left panel: with T large so that two stable steady solutions exist. Right panel: with T small so that the only steady solution corresponds to reentry.

waves on a ring have served well as a model for the study of arrhythmias. In that context, the stability of waves on a ring was studied in [**IG92**] [**CKG93**] [**CKG96**].

A simple model for a reentrant wave on a ring takes the form of an integral delay equation. We assume that the time of arrival of an action potential at position x on the ring is $t(x)$ so that at each position x

$$APD(x) + RT(x) = t(x) - t(x - L), \tag{4.3}$$

for a ring of length L. As above, we assume that the action potential duration is a function of the previous recovery time, $APD(x) = a(RT(x-L))$. (Note that on a ring of length L, $t(x-L)$ is the previous time of firing at position x.) We also assume that the local speed of travel is a function of the local recovery time,

$$C(x) = c(RT(x)). \tag{4.4}$$

Now we use that

$$t(x) - t(x-L) = \int_{x-L}^{x} \frac{dt}{dx} dx = \int_{x-L}^{x} \frac{1}{c(RT(x))} dx \tag{4.5}$$

to find the integral delay equation for $RT(x)$,

$$RT(x) = \int_{x-L}^{x} \frac{1}{c(RT(x))} dx - a(RT(x-L)). \tag{4.6}$$

In [**CKG93**][**CKG96**] it was shown that solutions of this equation are unstable if $a' > 1$. That this is the same as the condition for alternans in (3.13) is intriguing and has led to similar conjectures for higher dimensional arrhythmias, as discussed below. It is important to note, however, that the function a is an idealization and there is no true restitution curve for detailed ionic models. Thus, characterizations of stability based on properties of the restitution curve must be recognized as approximate.

It is important for clinical reasons to be able to ascertain if a reentrant pattern follows a one-dimensional path. This is usually done clinically using point stimuli from catheters inserted into the right atrium and examining the resulting entrainment patterns. A theoretical study of the response of a reentrant excitation on a ring to single stimuli is given in [**FR88**] [**GJ95**] [**NG96**], emphasizing the effect of pacing using point stimuli, entrainment patterns and annihilation.

5. Two-Dimensional Arrhythmias

The two most noticeable common reentrant arrhythmias are tachycardia and fibrillation . Both of these can occur on the atria (atrial tachycardia or flutter and atrial fibrillation) or on the ventricles (ventricular tachycardia and ventricular fibrillation). When they occur on the atria, they are not immediately life threatening because there is little disruption of blood flow. However, when they occur on the ventricles, they are life threatening. Ventricular fibrillation is fatal if it is not terminated quickly.

Tachycardia is often classified as being either monomorphic or polymorphic, depending on the assumed morphology of the activation pattern. Monomorphic tachycardia is identified as having a simple periodic ECG, while polymorphic tachycardia is usually quasiperiodic, apparently the superposition of more than one periodic oscillation. A typical example of a polymorphic tachycardia is called *Torsade de Pointe*, and appears on the ECG as a rapid oscillation with slowly varying amplitude (Fig. 12).

Two-dimensional arrhythmias are associated with two-dimensional self-sustained waves of activity, spiral waves, such as occur on the atria during atrial flutter or fibrillation. The reason these arrhythmias are thought of as two rather than three-dimensional is because the atrial wall is quite thin.

The first suggestion that spiral activity could be associated with arrhythmias came from the work of Wiener and Rosenbluth [**WR46**], although in their view a

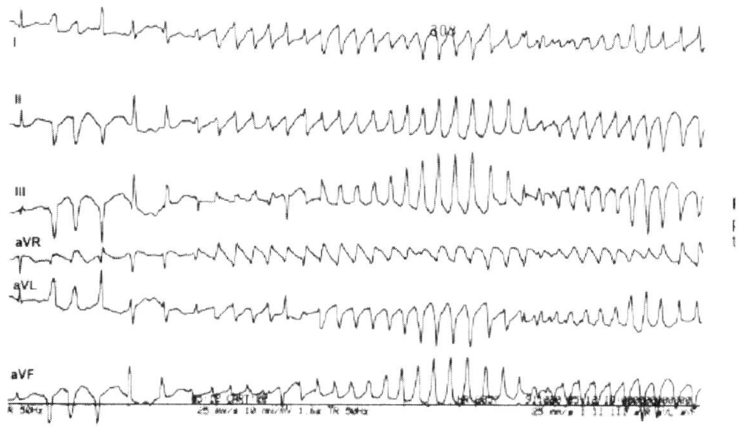

FIGURE 12. A six-lead ECG recording of Torsade de Pointe. From [**ZJ95**] Fig. 79-1, pg. 886.

spiral had to circulate around a nonfunctional hole, similar to the preparation of Mines with circulation around the superior vena cava.

A simple model for such a spiral is to assume, as above, that the normal velocity of waves is determined by their period. It is an easy matter to show that a wave with constant normal velocity is the involute of a circle, and is asymptotic to an Archimedean spiral [**KS98**], pg. 306. Thus, the period of rotation is related to the circumference of the hole around which the involute is rotating.

We now know that a nonfunctional hole at the center is not necessary for spiral activity. A less simple model for the evolution of a spiral with no hole is provided by the eikonal-curvature equation,

$$R_t \cdot \mathbf{n} = c - D\kappa \tag{5.1}$$

where $R_t \cdot \mathbf{n}$ is the normal velocity of the wave, c is the local plane wave velocity, a function of period, κ is the curvature of the wave, and D is a coupling (diffusion) coefficient. With this assumption on wave motion (which can be justified using asymptotic analysis [**Kee86**]), it is found that spirals are uniquely determined by a nonlinear eigenvalue problem, and they do not require a hole at their center [**TK88**]. Following this understanding, the mathematical study of spiral activity in models of two-dimensional excitable tissues has burgeoned [**Kee88b**] [**PS91**] [**Ber91**][**Kee92**][**Kar92**] [**KK96**].

Early numerical understandings of spiral activity were provided by finite state automata [**MRA64**] [**Kri66**] [**Kri78**] (see also [**SC84**]). The paradigm for initiation and maintenance of spiral waves suggested by [**MRA64**] was that they resulted from a high degree of random dispersion of refractoriness. This idea held sway for many years, but is gradually being recognized as inadequate.

The first experimental confirmation that reentrant waves could have a truly two-dimensional structure came from the mapping studies of Allessie et al. [**ABS73**] [**ABS76**], showing a wave of electrical activity rotating around a fixed point. It took much longer for cardiologists than mathematicians to recognize that reentrant activity was not necessarily one-dimensional, but could be truly two-dimensional [**JD93**] [**PDS$^+$93**]. For a long time in the cardiology literature, a reentrant arrhythmia was described as following a fixed reentrant conduction pathway, implicitly assumed to be one-dimensional. That this is incorrect is important to understand. An important difference between one-dimensional and two- or three-dimensional arrhythmias is that one-dimensional arrhythmias can *always* be eliminated by proper application of a point stimulus, while two- or three-dimensional arrhythmias cannot.

In the meantime, early numerical simulations of spiral activity using realistic ionic models showed that spirals could be unstable [**CW91**], and this led to the study of spiral stability [**Win90**] [**Bar94**] [**QWG00**]. That instabilities may lead to spiral breakup is the subject of numerous recent investigations [**IG91**] [**BE93**] [**Kar93**] [**PH93**] [**Kar94**] [**Pan98**], and remains the focus of ongoing research. In particular, it is of interest to determine the relationship between the restitution curve, alternans, spiral stability and breakup. The suggestion that alternans may be related to spiral breakup was first made in [**PH93**]. In [**Kar93**] [**Kar94**], Karma found that by varying certain parameters in a simple two-variable model, the period of spiral rotation could be reduced until alternans occurred. With further parameter variation, the amplitude of alternans increased until spiral breakup was observed. Similar observations have been made using a more detailed ionic model [**QWG99**] (see also [**RKG99**] [**GKV$^+$00**]), although there is as yet no mathematical analysis showing the relationship between a steep restitution curve and spiral stability and breakup.

Atrial arrhythmias are not immediately life threatening and so they can be allowed to continue in a patient for a long time. Of course, this is not particularly pleasant for a patient as it may require a decrease from normal activity. Long lasting atrial arrhythmias lead to the formation of clots in the atria, which if they get into the blood stream, can lead to strokes. Furthermore, atrial arrhythmias lead to atrial remodeling, which means that the longer an arrhythmia is allowed to proceed, the easier it becomes for a new arrhythmia to be initiated [**WKDA95**] [**LT97**]. The mechanism of remodeling and the propensity for reoccurrence is not understood.

The standard clinical method to eliminate two-dimensional atrial reentry is by application of a large (3-15 Joules) electrical shock. If this is done while the patient is conscious, it is quite painful. Thus, the search for efficient but less painful ways to eliminate reentrant atrial arrhythmias is of major current interest.

There are several ideas that are currently being pursued. It is known that some reentrant arrhythmias drift [**RP83**] [**DPS$^+$91**] because of tissue inhomogeneities. Spirals have also been observed to drift when they are periodically stimulated. Thus, the response of spirals to periodic forcing has become a topic of interest, with the hope that they can be forced to drift off of their finite spatial domain and thereby disappear [**SZM93**] [**BH94**] [**GZM95**] [**ZMM97**] [**NBH98**] [**PMZK00**]. To date, the clinical application of this idea has not found much success.

6. Three-Dimensional Arrhythmias

Three-dimensional arrhythmias are self-sustained waves of electrical activity that occur in the ventricles, and are called scroll waves . They are three dimensional, and not two-dimensional, because the ventricular wall is quite thick. The

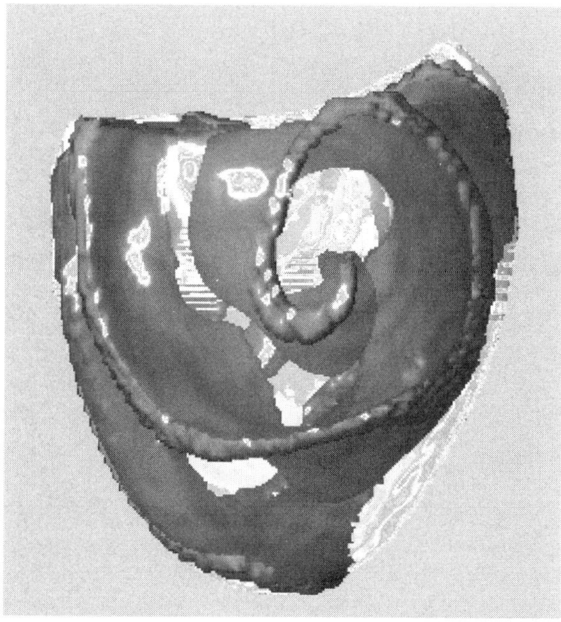

FIGURE 13. Numerically computed scroll wave in ventricular muscle. From [**PK95b**], Fig. 3a. p. 685.

observation that reentrant waves in the ventricles had to be scroll-like came rather early from the work of Winfree [**Win73**], however, the experimental verification of this did not come until much later [**FKC+88**].

The simplest example of a reentrant ventricular arrhythmia, monomorphic tachycardia, is probably a scroll wave [**Win87**], however, ventricular fibrillation certainly has a more complicated three-dimensional structure and behavior [**WP90**] [**WLP+98**].

Our current understanding of scroll waves comes from two sources, mathematical analysis of greatly idealized models, and numerical simulation of less idealized models.

The mathematical analysis of scroll waves has focused on the behavior of the organizing center of the scroll, the scroll wave filament. In a series of papers [**WS84a**] [**WS83a**] [**WS83b**] [**WS83c**] [**WS84b**], Winfree and Strogatz explored the topology of scroll wave filaments. General equations of motion for scroll wave filaments were derived in [**Kee88a**], and the consequences of these equations explored in [**Bik89**] [**KT90**] [**KT92**] [**BHZ94**].

The study of the two- and three- dimensional arrhythmias is complicated by the fact that cardiac tissue is an anisotropic, inhomogeneous, irregularly shaped, bidomain [**KS98**]. At this point in time, our understanding of these waves for realistic tissue models comes almost entirely from numerical simulations. A three-dimensional view of a (numerically computed) monomorphic V-tach is shown in Fig. 13.

Stable monomorphic ventricular tachycardia is rare, as most reentrant tachycardias are unstable and degenerate into fibrillation. It is possible that a stable

FIGURE 14. Surface view of fibrillatory reentrant activity in the ventricles (computed by A. V. Panfilov)

monomorphic tachycardia on the ventricles is actually a one-dimensional arrhythmia in that it passes through a small pathway in imperfectly infarcted tissue and this may suggest a strategy by which to eliminate them.

Fibrillation is believed to correspond to the presence of many reentrant patterns moving throughout the ventricles in continuous, perhaps erratic, fashion, leading to an uncoordinated pattern of ventricular contraction and relaxation. A surface view of a (numerically computed) fibrillatory pattern is shown in Fig. 14.

The likely reason that monomorphic V-tach is rare is because there are a number potential sources of instability, although the mechanisms of the instability have not been decisively determined. A leading contender for the cause of instability is the steep slope of the restitution curve. Until recently, however, there was no experimental evidence in support of this hypothesis, as data on restitution properties showed slopes that were consistently less than one [**ES83**]. This situation changed dramatically when steep restitution curves were found in dog myocardium [**KRGJ98**]. According to these authors, the previous findings were because restitution properties were measured at a relatively long periods of stimulation, where restitution curves can be expected to be quite flat. With their improved technique, they were able to measure restitution properties at much shorter cycle lengths, finding slopes as large as 1.7.

Additional evidence for the importance of the restitution curve comes from studies showing that certain drugs (diacetyl monoxin and verapamil in [**RKG99**] and bretylium in [**GKV$^+$00**]) can flatten the restitution curve and actually prevent the occurrence of ventricular fibrillation. In [**RKG99**] it was also found that another anti-arrhythmic drug, procainamide, did not reduce the slope of the restitution curve, and neither did it prevent induction of ventricular fibrillation. These experiments also provided a plausible explanation for why a breakup instability was not seen in earlier studies. The earlier studies used fluorescent dye techniques which required electrical-mechanical uncouplers to prevent motion artifacts in the optical recordings. The decouplers used were diacetyl monoxin and verapamil, which are now known to also flatten restitution curves and thereby make spiral breakup less likely.

A steep restitution curve is not likely to be the only cause of instability, as there are probably several other, mostly geometrical, factors. It is curious that breakup of spirals or scrolls occurs more readily in three-dimensions than in two [**PH95**] [**PH96**]. Some other possibilities of instability mechanisms include vortex shedding in the vicinity of spatial inhomogeneity [**PK93b**] [**AKMP91**] [**CPD+96**], scroll wave filament instability [**BHZ94**], rotational anisotropy of myocardial tissue [**PK95a**] [**PK93a**] [**PK95b**] [**FK98**] [**QXG99**] [**QKGW00**] and gradual tissue heterogeneities [**PJ95**].

The observation that reentrant arrhythmias on the ventricles are three-dimensional has important clinical consequences. As we have seen, if an arrhythmia is one-dimensional, then there must be a point somewhere on the circuit that can be destroyed that will eliminate the arrhythmia. Similarly, if a two-dimensional arrhythmia is the result of rapid periodic pacing from an ectopic focus (i.e., not reentry), then destroying the oscillator will eliminate the arrhythmia [**JHS+97**]. However, if a point on the surface of the ventricles appears to be a point of first activation, there is no guarantee that this point is an ectopic focus or a point on a one-dimensional circuit. In fact, a reentrant arrhythmia that is circulating deep within the tissue may have a point on the surface at which it first emerges, but because the arrhythmia is sustained by a three-dimensional reentrant cycle, destruction of the surface cells will have no effect whatever on the arrhythmia. This fact is unfortunately not completely appreciated by all cardiac electrophysiologists, and this misunderstanding has resulted in some disappointing surgical procedures.

7. Conclusion

Our understanding of cardiac arrhythmias is enhanced by viewing them from the vantage point of their spatial dimension. From this perspective, we not only learn some interesting mathematics, but we also gain important practical insights. In fact, with mathematical reasoning and models we are able to visualize behavior (for example, the three dimensional structure of ventricular tachycardia and fibrillation) that is far beyond the current capabilities of experimental techniques.

While many difficulties remain unresolved, there is reasonable hope that these mathematical ideas will continue to be fruitful.

References

[ABS73] M. A. Allessie, F. I. M. Bonke, and F. J. G. Schopman, *Circus movement in rabbit atrial muscle as a mechanism of tachycardia*, Circ. Res. **33** (1973), 54–62.

[ABS76] M. A. Allessie, F. I. M. Bonke, and F. J. G. Schopman, *Circus movement in rabbit atrial muscle as a mechanism of tachycardia.II.the role of nonuniform recovery of excitability in the occurrence of unidirectional block as studied with multiple microelectrodes*, Circ. Res. **39** (1976), 168–177.

[AKMP91] K.I. Agladze, J.P. Keener, S.C. Mueller, and A.V Panfilov, *Rotating spiral waves created by geometry*, Science **264** (1991), 1746–1748.

[AM80] C. Antzelevitch and G. K. Moe, *Characteristics of reflection as a mechanism of reentrant arrhythmias and its relationship to parasystole*, Circ. **61** (1980), 182–191.

[Bar94] D. Barkley, *Euclidean symmetry and the dynamics of rotating spiral waves*, Phys. Rev. Lett. **72** (1994), 164–167.

[BAZS94] J. Billette, F. Amellal, J. Zhao, and A. Shrier, *Relationship between different recovery curves representing rate-dependent AV nodal function in rabbit*, J. Electrovasc. Electrophysiol. **5** (1994), 63–75.

[BE93] M. Bar and M. Eiswirth, *Turbulence due to spiral breakup in a continuous excitable medium*, Phys. Rev. E. **48** (1993), 1635–1637.

[Ber91] A. Bernoff, *Spiral wave solutions for reaction-diffusion equations in a fast-reaction/slow-diffusion limit*, Phys. D **53** (1991), 125–150.

[BH94] V. N. Biktashev and A. V. Holden, *Design principles of a low voltage cardiac defibrillator based on the effect of feedback resonant drift*, J. theor. Biol **169** (1994), 101–112.

[BHZ94] V. N. Biktashev, A. V. Holden, and H. Zhang, *Tension of organizing filaments of scroll waves*, Phil. Trans. R. Soc. Lond. A **347** (1994), 611–630.

[Bik89] V.N. Biktashev, *Evolution of twist on autowave vortex*, Physica D **36** (1989), 467–472.

[BR77] G. W. Beeler and H. J. Reuter, *Reconstruction of the action potential of ventricular myocardial fibers*, J. Physiol. **268** (1977), 177–210.

[Car99] Edward Carmeliet, *Cardiac ionic currents and acute ischemia: From channels to arrhythmias*, Physiological Reviews **79** (1999), no. 3, 917–1017.

[CGJ90] D. R. Chialvo, R. F. Gilmour, Jr., and J. Jalife, *Low dimensional chaos in cardiac tissue*, Letters to Nature **343** (1990), 653–657.

[CGRG89] M. Courtemanche, L. Glass, M. D. Rosenbarten, and A. L. Goldberger, *Beyond pure parasystole: promises and problems in modeling complex arrhythmias*, Am. J. Physiol. **257** (1989), H693–706.

[CKG93] M. Courtemanche, J. P. Keener, and L. Glass, *Instabilities of a propagating pulse in a ring of excitable media*, Phys. Rev. Letts. **70** (1993), 2182–2185.

[CKG96] M. Courtemanche, J. P. Keener, and L. Glass, *A delay equation representation of pulse circulation on a ring in excitable media*, SIAM J. Appl. Math. **56** (1996), 119–142.

[CL84] E. A. Coddington and N. Levinson, *Ordinary differential equations*, Robert E. Krieger Publishing Company, 1984.

[CPD+96] C. Cabo, A.M. Pertsov, J.M. Davidenko, W.T. Baxter, R.A. Gray, and J. Jalife, *Vortex shedding as a precursor of turbulent electrical activity in cardiac muscle*, Biophys. J. **70** (1996), 1105–1111.

[CRN98] M. Courtemanche, R. J. Ramirez, and S. Nattel, *Ionic mechanism underlying human atrial action potential properties: insights from a mathematical model*, Am. J. Physiol. **275** (1998), H301–H321.

[CVJCN98] F. F. T. Ch'en, R. D. Vaughan-Jones, K. Clarke, and D. Noble, *Modelling myocardial ischaemia and reperfusion*, Prog. Biophys. Mol. Biol. **69** (1998), 515–538.

[CW91] M. Courtemanche and A. T. Winfree, *Re-entrant rotating waves in a Beeler-Reuter-based model of 2-dimensional cardiac electrical activity.*, Int. J. Bif. and Chaos **1** (1991), 431–444.

[DN85] D. DiFrancesco and D. Noble, *A model of cardiac electrical activity incorporating ionic pumps and concentration changes*, Phil. Trans. R. Soc. B **307** (1985), 353–398.

[DPS+91] J. M. Davidenko, A. V. Pertsov, R. Salomonsz, W. Baxter, and J. Jalife, *Stationary and drifting spiral waves of excitation in isolated cardiac muscle*, Nature **355** (1991), 349–351.

[ES83] V. Elharrar and B. Surawicz, *Cycle length effect on restitution of action potential duration in dog cardiac fibers*, Am. J. Physiol. **244** (1983), H782–H792.

[FK98] F. Fenton and A. Karma, *Vortex dynamics in three-dimensional continuous myocardium with fiber rotation: filament instability and fibrillation*, Chaos **8** (1998), 20–47.

[FKC+88] D. W. Frazier, W. Krassowska, P. S. Chen, P. D. Wolf, N. D. Danieley, W. M. Smith, and R. E. Ideker, *Transmural activations and stimulus potentials in three-dimensional anisotropic canine myocardium*, Circ. Res. **63** (1988), 135–146.

[FR88] L. H. Frame and E. K. Rhee, *Reversal of reentry by pacing: relation to termination*, Circ. **80** (1988), 11–96.

[FSFT96] J. M. Ferrero, J. Sa'iz, J. M. Ferrero, and N. V. Thakor, *Simulation of action potential from metabolically impaired cardiac myocytes*, Circ. Res. **79** (1996), no. 2, 208–221.

[FSS+83] M.R. Franz, J. Schaefer, M. Schóttler, W.A. Seed, and M.I.M. Noble, *Electrical and mechanical restitution of the human heart at different rates of stimulation*, Circ. Res. **53** (1983), 815–822.

[GG82] M. R. Guevara and L. Glass, *Phase locking, period doubling bifurcations and chaos in a mathematical model of a periodically driven oscillator: a theory for the entrainment of biological oscillators and the generation of cardiac dysrhythmias*, J. Math. Biol. **14** (1982), 1–23.

[GG99] G. Gedeon and L. Glass, *Continuity of resetting curves for FitzHugh-Nagumo equations on a circle*, Fields Institute Comms. **21** (1999), 225–236.

[GGB86] L. Glass, A. L. Goldberger, and J. Bélair, *Dynamics of pure parasystole*, Am. J. Physiol. **251** (1986), H841–H847.

[GGBS84] L. Glass, M. R. Guevara, J. Bélair, and A. Shrier, *Global bifurcations of a periodically forced biological oscillator*, Physical Review A **29** (1984), no. 3, 1348–1357.

[GGCS87] L. Glass, A. L. Goldberger, M. Courtemanche, and A. Shrier, *Nonlinear dynamics, chaos and complex cardiac arrhythmias*, Proc. R. Soc. Lond. A **413** (1987), 9–26.

[GGSP83] L. Glass, M. R. Guevara, A. Shrier, and R. Perez, *Bifurcation and chaos in a periodically stimulated cardiac oscillator*, Physica D **7** (1983), 89–101.

[GJ95] L. Glass and M. E. Josephson, *Resetting and annihilation of reentrant abnormally rapid heartbeat*, Phys. Rev. Letts **75** (1995), 2059–2062.

[GK95] L. Glass and D. Kaplan, *Understanding nonlinear dynamics*, Springer-Verlag, New York, 1995.

[GKV+00] A. Garfinkel, Y.H. Kim, O. Voroshilovsky, Z. Qu, J.R. Kil, M.H. Lee, H.S. Karagueuzian, J.N. Weiss, and P.S. Chen, *Preventing ventricular fibrillation by flattening cardiac restitution*, Proc. Natl. Acad. Sci. USA **97** (2000), 6061–6066.

[GM79] L. Glass and M. C. Mackey, *A simple model for phase locking of biological oscillators*, J. Math. Biol. **7** (1979), 339–352.

[GM88] L. Glass and M. C. Mackey, *From clocks to chaos: The rhythms of life*, Princeton University Press, 1988.

[GP82] L. Glass and R. Perez, *Fine structure of phase locking*, Physical Review Letters **48** (1982), no. 26, 1772–1775.

[GS94] L. Glass and J. Sun, *Periodic forcing of a limit-cycle oscillator: Fixed points, Arnold tongues, and the global organization of bifurcations*, Physical Review E **50** (1994), no. 6, 5077–5084.

[GZM95] S. Grill, V. S. Zykov, and S. C. Müller, *Feedback-controlled dynamics of meandering spiral waves*, Phys. Rev. Lett. **75** (1995), 3368–3371.

[HH52] A. L. Hodgkin and A. F. Huxley, *A quantitative description of membrane current and its application to conduction and excitation in nerve*, J. Physiol. **117** (1952), 500–544.

[IG91] H. Ito and L. Glass, *Spiral breakup in a new model of discrete excitable media*, Phys. Rev. Lett. **66** (1991), 671–674.

[IG92] H. Ito and L. Glass, *Theory of reentrant excitation in a ring of cardiac tissue*, Physica D **56** (1992), 84–106.

[JA79] J. Jalife and C. Antzelevitch, *Phase resetting and annihilation of pacemaker activity in cardiac tissue*, Science **206** (1979), 695–697.

[JD93] J. Jalife and J. M Davidenko, *Spiral waves as a mechanism of reentrant excitation in isolated cardiac muscle*, Cardiac mapping (Mount Kisko,NY) (M. Shenasta, M. Borggrefe, and Breithardt G., eds.), Futura, 1993, pp. 607–623.

[JHS+97] P. Jais, M. Haissaguerre, D. C. Shah, S. Chouairi, L. Gencel, M. Hocini, and J. Clementy, *A focal source of atrial fibrillation treated by discrete radiofrequency ablation*, Circulation **95** (1997), 572–576.

[JML86] J. Jalife, D. C. Michaels, and R. Langendorf, *Modulated parasystole originating in the sinoatrial node*, Circulation **74** (1986), no. 5, 945–954.

[JRW98] M. S. Jafri, J. J. Rice, and R. L. Winslow, *Cardiac Ca^{2+} dynamics: The roles of ryanodine receptor adaptation and sarcoplasmic reticulum load*, Biophysical Journal **74** (1998), 1149–1168.

[Kar92] A. Karma, *Scaling regime of spiral wave propagation in single-diffusive media*, Phys. Rev. Lett. **68** (1992), 397.

[Kar93] A. Karma, *Spiral breakup in model equations of action potential propagation in cardiac tissue*, Phys. Rev. Lett. **71** (1993), 1103–1106.

[Kar94] A. Karma, *Electrical alternans and spiral wave breakup in cardiac tissue*, Chaos **4** (1994), 461–472.

[Kee80] J.P. Keener, *Chaotic behavior in piecewise continuous difference equations*, Trans. AMS **261** (1980), 589–604.

[Kee81] J.P. Keener, *On cardiac arrhythmias: Av conduction block*, J. Math. Biol. **12** (1981), 215–225.

[Kee84] J. P. Keener, *Dynamic patterns in excitable media*, Lecture Notes in Biomathematics, vol. 55, pp. 157–169, Springer Verlag, 1984.

[Kee86] J. P. Keener, *A geometrical theory for spiral waves in excitable media*, SIAM J. Appl. Math. **46** (1986), 1039–1056.

[Kee87] J. P. Keener, *Causes of propagation failure in excitable media*, Temporal Disorder in Human Oscillatory Systems (L. Rensing, U. an der Heiden, and M. C. Mackey, eds.), Springer Verlag, Berlin, 1987.

[Kee88a] J. P. Keener, *The dynamics of three dimensional scroll waves in excitable media*, Physica D **31** (1988), 269–276.

[Kee88b] J. P. Keener, *On the formation of circulating patterns of excitation in anisotropic excitable media*, J. Math. Biol. **26** (1988), 41–56.

[Kee92] J. P. Keener, *The core of the spiral*, SIAM J. Appl. Math. **52** (1992), 1372–1390.

[KG84] J.P. Keener and L. Glass, *Global bifurcations of a periodically forced oscillator*, J. Math. Biol. **21** (1984), 175–190.

[KHR81] J. P. Keener, F. C. Hoppensteadt, and J. Rinzel, *Integrate and fire models of nerve membrane response to oscillatory input*, SIAM J. Appl. Math. **41** (1981), 503–517.

[KK96] D. A. Kessler and R. Kupferman, *Spirals in excitable media. ii: Meandering transition in the diffusive free-boundary limit*, Phys. D (1996).

[Kni72] B. W. Knight, *Dynamics of encoding a population of neurons*, J. Gen. Physiol. **59** (1972), 734–766.

[KRGJ98] M.L. Koller, M.L. Riccio, and R.F. Gilmour Jr, *Dynamics restitution of action potential duration during electrical alternans and ventricular fibrillation*, Am.J.Physiol **275** (1998), H1635–H1642.

[Kri66] V. I. Krinsky, *Spread of excitation in an inhomogeneous medium (state similar to cardiac fibrillation)*, Biophysics **11** (1966), 776–784.

[Kri78] V. I. Krinsky, *Mathematical models of cardiac arrythmias (spiral waves)*, Pharm. Ther. B. **3** (1978), 539–555.

[KS98] J. Keener and J. Sneyd, *Mathematical Physiology*, Springer-Verlag, New York, 1998.

[KT90] J. P. Keener and J. J. Tyson, *Helical and circular scroll wave filaments*, Physica D **44** (1990), 191– 202.

[KT92] J. P. Keener and J. J. Tyson, *The dynamics of scroll waves in excitable media*, SIAM Review **34** (1992), 1–39.

[LG91] M. Lewis and P. Grindrod, *One-way block in cardiac tissue: a mechanism for propagation failure in Purkinje fibres*, Bull. Math. Biol. **53** (1991), 881–899.

[LGBP92] J. E. Lewis, L. Glass, M. Bachoo, and Canio Polosa, *Phase resetting and fixed-delay stimulation of a simple model of respiratory rhythm generation*, J Theor. Biol. **259** (1992), 491–506.

[LR94a] C. H. Luo and Y. Rudy, *A dynamic model of the cardiac ventricular action potential; I: Simulations of ionic currents and concentration changes*, Circ. Res. **74** (1994), 1071–1096.

[LR94b] C. H. Luo and Y. Rudy, *A dynamic model of the cardiac ventricular action potential; II: Afterdepolarizations, triggered activity and potentiation*, Circ. Res. **74** (1994), 1097–1113.

[LT97] Chu-Pak Lau and Hung-Fat Tse, *Electrical remodelling of chronic atrial fibrillation*, Clinical and Experimental Pharmacology and Physiology **24** (1997), 982–983, from

Proceedings of the Symposium on Cardiovascular Science and Medicine: From Bench to Bedside.

[MFV88] O. Piro M. Feingold, D. L. Gonzalez and H. Viturro, *Phase locking, period doubling, and chaotic phenomena in externally driven excitable systems.*, Phys. Rev. A **37** (1988), 4060–4063.

[Min14] G. R. Mines, *On circulating excitations in heart muscle and their possible relation to tachycardia and fibrillation*, Trans. Roy. Soc. Can. **4** (1914), 43–53.

[MJMM77] G. K. Moe, J. Jalife, W. J. Mueller, and B. Moe, *A mathematical model of parasystole and its application to clinical arrhythmias*, Circulation **56** (1977), no. 6, 968.

[MR81] R. Miller and J. Rinzel, *The dependence of impulse propagation speed on firing frequency, dispersion for the Hodgkin-Huxley models*, Biophys. J. **34** (1981), 227–259.

[MRA64] G. K. Moe, W. C. Rheinbolt, and J. A. Abildskov, *A computer model of atrial fibrillation*, Am. Heart J. **67** (1964), 200–220.

[NBH98] E.V. Nikolaev, V. N. Biktashev, and A. V. Holden, *On feedback resonant drift and interaction with the boundaries in circular and annular excitable media*, Chaos, Solitons and Fractals **9** (1998), 363–376.

[NG96] T. Nomura and L. Glass, *Entrainment and termination of reentrant wave propagation in a periodically stimulated ring of excitable media*, Phys. Rev. **53** (1996), 6353–6360.

[Nob62] D. Noble, *A modification of the Hodgkin-Huxley equations applicable to Purkinje fiber action and pacemaker potential*, J. Physiol. **160** (1962), 317–352.

[Pan98] A. V. Panfilov, *Spiral breakup as a model of ventricular fibrillation*, Chaos **8** (1998).

[Pau82] J. P. Pauwelussen, *One way traffic of pulses in a neuron*, J. Math. Biol. **15** (1982), 151–171.

[PDS+93] A. M. Pertsov, J. M. Davidenko, R. Salomontsz, W. Baxter, and J. Jalife, *Spiral waves of excitation underlie reentrant activity in isolated cardiac muscle*, Circ. Res. **72** (1993), 631–650.

[PH93] A. V. Panfilov and P. Hogeweg, *Spiral break-up in a modified FitzHugh–Nagumo model*, Phys. Lett. A **176** (1993), 295–299.

[PH95] A.V. Panfilov and P. Hogeweg, *Turbulence in a three-dimensional excitable media*, Science **270** (1995), 1223–1224.

[PH96] A. V. Panfilov and P. Hogeweg, *Turbulence in a three-dimensional excitable medium*, Phys. Rev. E **53** (1996), 1740–1743.

[PJ95] A. M. Pertsov and J. Jalife, *Scroll waves in three-dimensional cardiac muscle*, Cardiac electrophysiology. From cell to bedside, 3rd edition (Philadelphia) (D. P. Zipes and J. Jalife, eds.), Saunders, 1995, pp. 336–344.

[PK93a] A. V. Panfilov and J. P. Keener, *Generation of reentry in anisotropic myocardium*, J. Card. Electrophys. **4** (1993), 412–421.

[PK93b] A.V. Panfilov and J.P. Keener, *Effects of high frequency stimulation in excitable medium with obstacle*, J. Theor. Biol. **163** (1993), 439–448.

[PK95a] A. V. Panfilov and J. P. Keener, *Re-entry in an anatomical model of the heart*, Chaos, Solitons and Fractals **5** (1995), 681–689.

[PK95b] A. V. Panfilov and J. P. Keener, *Reentry in 3-dimensional Fitzhugh-Nagumo medium with rotational anisotropy*, Physica D **84** (1995), 545–552.

[PMZK00] A. V. Panfilov, S. C. Muller, V. S. Zykov, and J. P. Keener, *Defibrillation of cardiac tissue by multiple subthreshold shocks*, Phys. Rev. E **61** (2000), 4644–4647.

[PS91] P. Pelce and J. Sun, *Wave fronts interaction in steadily rotating spirals*, Phys. D **48** (1991), 353–366.

[QKGW00] Z. Qu, J. Kil, A. Garfinkel, and J. N. Weiss, *Scroll wave dynamics in a three-dimensional cardiac tissue model: Role of restitution, thickness, and fiber rotation*, Biophys. J. **78** (2000), 2761–2775.

[QWG99] Z. Qu, J.N. Weiss, and A. Garfinkel, *Cardiac electrical restitution properties and stability of reentrant spiral waves: a simulation study*, Am. J. Physiol. **276** (1999), H269–H283.

[QWG00] Z. Qu, J. N. Weiss, and A. Garfinkel, *From local to global spatiotemporal chaos in a cardiac tissue model*, Phys. Rev. E **61** (2000), no. 1, 727–732.

[QXG99] Z. Qu, F. Xie, and A. Garfinkel, *Diffusion-induced vortex filament instability in 3-dimensional excitable media*, Physical Review Letters **83** (1999), no. 13, 2668–2671.

[RJW99] J. J. Rice, M. S. Jafri, and R. L. Winslow, *Modeling gain and gradedness of* Ca^{2+} *release in the functional unit of the cardiac diadic space*, Biophysical Journal **77** (1999), no. 4, 1871–1884.

[RKG99] M. L. Riccio, M. L. Koller, and R. F. Gilmour, Jr., *Electrical restitution and spatiotemporal organization during ventricular fibrillation*, Circ Res. **84** (1999), 955–963.

[RP83] A.N. Rudenko and A.V. Panfilov, *Drift and interaction of vortices in two-dimensional heterogeneous active medium*, Studia Biophysica **98** (1983), 183–188.

[Ru76] R.F. Rushmer, *Structure and Function of the Cardiovascular System (second edition)*, W.B. Saunders Co., Philadelphia, PA, 1976.

[SAGB95] J. Sun, F. Amellal, L. Glass, and J. Billette, *Alternans and period-doubling bifurcations in atrioventricular nodal conduction*, J. Theor. Biol. **173** (1995), 79–91.

[SC84] J.M. Smith and R.J. Cohen, *Simple finite element model accounts for wide range of cardiac dysrhythmias*, Proc. Natl. Acad. Sci. USA **81** (1984), 233–237.

[Str94] S. H. Strogatz, *Nonlinear dynamics and chaos*, Addison-Wesley, Reading, Massachusetts, 1994.

[SZM93] O. Steibock, V. S. Zykov, and S. C. Müller, *Control of spiral-wave dynamics in active media by periodic modulation of excitability*, Nature **366** (1993), 322–324.

[TK88] J. J. Tyson and J. P. Keener, *Singular perturbation theory of traveling waves in excitable media*, Physica D **32** (1988), 327–361.

[vdPvdM28] B. van der Pol and J. van der Mark, *The heartbeat considered as a relaxation oscillation, and an electrical model of the heart*, Phil. Mag. **6** (1928), 763–775.

[Wen04] K. F. Wenckebach, *Arrhythmia of the heart: a physiological and clinical study*, Green, Edinburgh, 1904.

[Win73] A. T. Winfree, *Scroll-shaped waves of chemical activity in three dimension*, Science **181** (1973), 937–939.

[Win87] A. T. Winfree, *When time breaks down*, Princeton University Press, Princeton, 1987.

[Win90] A. T. Winfree, *Varieties of spiral wave behavior: An experimental approach to the theory of excitable media*, Chaos **1** (1990), 303–334.

[WKDA95] M. C. E. F. Wijffels, Ch. J. H. J. Kirchhof, R. Dorland, and M. A. Allessie, *Atrical fibrillation begets atrical fibrillation. a study in awake chronically instrumented goats*, Circulation **92** (1995), 1954–1968.

[WKV+93] R. L. Winslow, A. Kimball, T. Varghese, C. Adlakha, and D. Noble, *Generation and propagation of ectopic beats induced by Na-K pump inhibition in atrial network model*, Proc. R. Soc. Lond. B. **254(1339)** (1993), 55–61.

[WKVN93] R. L. Winslow, A. Kimball, T. Varghese, and D. Noble, *Simulating cardiac sinus and atrial network dynamics on the Connection Machine*, Physica D **68** (1993), 364–386.

[WLP+98] F. X. Witkowski, L. J. Leon, P. A. Penkoske, W. R. Giles, M. L. Spano, W. L. Ditto, and A. T. Winfree, *Spatiotemporal evolution of ventricular fibrillation*, Nature **392** (1998), 78–82.

[WOG95] M. Watanabe, N. F. Otani, and R. F. Gilmour, Jr., *Biphasic restitution of action potential duration and complex dynamics in ventricular myocardium*, Circ Res. **76** (1995), 515–921.

[WP90] F. X. Witkowski and P. A. Penkoske, *Activation patterns during ventricular fibrillation*, Mathematical approaches to cardiac arrhythmias. Annals of the New York Academy of Sciences (J. Jalife, ed.), New York Academy of Sciences, 1990, pp. 219–231.

[WR46] N. Wiener and A. Rosenbluth, *The mathematical formulation of the problem of conduction of impulses in a network of connected excitable elements, specifically in cardiac muscle*, Arch. Inst. Cardiol. Mex. **16** (1946), 205–265.

[WS83a] A. T. Winfree and S. H. Strogatz, *Singular filaments organize chemical waves in three dimensions: 1. geometrically simple waves.*, Phys. D **8** (1983), 35–49.

[WS83b] A. T. Winfree and S. H. Strogatz, *Singular filaments organize chemical waves in three dimensions: 2. twisted waves.*, Phys. D **9** (1983), 65–80.

[WS83c] A. T. Winfree and S. H. Strogatz, *Singular filaments organize chemical waves in three dimensions: 3. knotted waves.*, Phys. D **9** (1983), 333–345.

[WS84a] A. T. Winfree and S. H. Strogatz, *Organizing centers for three-dimensional chemical waves*, Nature **311** (1984), 611–615.

[WS84b] A. T. Winfree and S. H. Strogatz, *Singular filaments organize chemical waves in three dimensions: 4. wave taxonomy.*, Phys. D **13** (1984), 221–233.

[YJAG99] Ali R. Yehia, Dominique Jeandupeux, Francisco Alonso, and Michael R. Guevara, *Hysteresis and bistability in the direct transition from 1:1 to 2:1 rhythm in periodically driven single ventricular cells*, Chaos **9** (1999), no. 4, 916–931.

[YNI80] K. Yanagihara, A. Noma, and H. Irisawa, *Reconstruction of sino-atrial node pacemaker potential based on voltage clamp experiments*, Jap. J. Physiol. **30** (1980), 841–857.

[ZJ95] D. P. Zipes and J. Jalife (eds.), *Cardiac electrophysiology. from cell to bedside, 2nd edition*, Philadelphia, Saunders, 1995.

[ZMM97] V. S. Zykov, A. S. Mikhailov, and S. C. Müller, *Controlling spiral waves in confined geometries by global feedback*, Phys. Rev. Lett. **78** (1997), 3398–3401.

Department of Mathematics, University of Utah, Salt Lake City, UT 84112
E-mail address: keener@math.utah.edu

Proceedings of Symposia in Applied Mathematics
Volume **59**, 2002

Calcium Excitability

James Sneyd

ABSTRACT. In some of the other talks we have seen how active control of the membrane potential leads to complicated dynamical behavior, and how cells can thus use their membrane potential as a intercellular signalling mechanism.

What is not so well-known is the fact that calcium plays a similar role as a messenger, both within single cells and between multiple cells. The study of the dynamic properties of calcium shows many mathematical similarities with models based on electrical excitability, but the physiological mechanisms are quite distinct.

In response to agonists such as hormones or neurotransmitters, many cell types exhibit oscillations in the concentration of intracellular free calcium ions. In larger cells these oscillations often take the form of periodic intracellular waves, and such waves are commonly seen traveling from cell to cell to form an intercellular calcium wave. The exact physiological function of these calcium waves and oscillations is not completely clear in any cell type. Nevertheless, they are believed to be one important mechanism whereby cells can control their behavior, and coordinate with their neighbors.

Not only is the study of calcium oscillations and waves important from the physiological point of view, the mathematics involved is non-trivial and interesting. Most models of calcium waves are of the reaction-diffusion type, and, as yet, very little is known about the properties of traveling wave solutions in these highly non-linear systems. The existence and importance of calcium buffers, and the effects of intercellular boundaries, add additional complications to what is already a difficult problem.

Here I present a brief history of models of calcium waves and oscillations, from the early model of Goldbeter, Dupont and Berridge in 1990, through to more recent detailed models. I shall then discuss how the modelling work has answered (at least partially) specific physiological questions, and thus show how mathematicians and experimentalists can combine effectively to address problems that would not be soluble by either working independently. I shall end by discussing some of the mathematical questions raised by this modelling work, and present a number of open problems.

1991 *Mathematics Subject Classification.* 35K57, 92C05, 92C30, 92C37.

Key words and phrases. Calcium oscillations, calcium waves, cellular physiology, reaction-diffusion equations.

This work was supported by the Marsden Fund of the Royal Society of New Zealand.

1. Cellular calcium homeostasis

In the previous chapters we have seen the importance of cellular electrical excitability. As a by-product of the mechanisms used to control its volume, every cell maintains a potential difference across its cell membrane (Keener and Sneyd, 1998). In cells such as neurons or muscle cells this membrane potential difference is used as a signalling device, allowing one cell to communicate with its neighbors, or stimulating such things as muscular contraction or hormonal secretion. Studies of the membrane potential have been an important area of applied mathematics for many years, and the mathematical literature on cellular electrophysiology is now vast. For instance, in this volume we see models of large numbers of neurons hooked together to form neural networks as in the chapter by Terman, models of specialized neurons for the transformation of a light signal into an electrical signal as in the chapter by Tranchina, and the behavior of spreading waves of electrical activity in cardiac muscle tissue as in the chapter by Keener. Detailed descriptions of the physiology and mathematics of cellular electrophysiology can be found in Berne and Levy (1993), Guyton and Hall (1996), Keener and Sneyd (1998) and Johnston and Wu (1995).

What is not so well known is that calcium ions play a similar role in both intracellular and intercellular signalling, and show many of the same dynamic behaviors as have been discovered in electrophysiology. In the last ten to fifteen years it has become clear that the concentration of free intracellular calcium ions is controlled by a dynamical system every bit as complex as that which regulates the membrane potential, and one that shows the same features of oscillatory responses, excitability and traveling waves. Thus the study of the dynamics of calcium homeostasis has developed into a very exciting field for applied mathematicians.

Calcium is critically important for a vast array of cellular functions as can be seen by a quick look through any physiology book. Thus, the mechanisms by which a cell controls its calcium concentration are of central interest in cell physiology. There are a number of calcium control mechanisms operating on different levels, all designed to ensure that calcium is present in sufficient quantity to perform its necessary functions, but not in too great a quantity in the wrong places. Prolonged high concentrations of calcium are toxic. For example, since calcium causes contraction of muscle cells, failure to remove calcium can keep a muscle cell in a state of constant tension (as in rigor mortis).

In vertebrates, the majority of body calcium is stored in the bones, whence it can be released by hormonal stimulation to maintain an extracellular calcium concentration of around 1 mM, while the intracellular calcium concentration is kept at around 0.1 μM. Since the internal concentration is low, there is a steep concentration gradient from the outside of a cell to the inside. This disparity has the advantage that cells are able to raise their calcium concentration quickly, by opening calcium channels and relying on passive flow down a steep concentration gradient, but has the disadvantage that energy must be expended to keep cytosolic calcium concentration low. Thus, cells have finely tuned mechanisms to control the influx and removal of cytosolic calcium.

Calcium is removed from the cytoplasm in two principal ways — it is pumped out of a cell, and it is sequestered into internal membrane-bound compartments such as the mitochondria, the endoplasmic reticulum (ER) or sarcoplasmic reticulum (SR), and secretory granules (Fig. 1). Since the calcium concentration in the cytoplasm is much lower than either the extracellular concentration or the concentration inside the internal compartments, both methods of calcium removal require expenditure of energy. Some of this is by a calcium ATPase that uses energy stored in ATP to pump calcium out of the cell, or into an internal compartment. There is also a Na^+-Ca^{2+} exchanger in the cell membrane that uses the energy of the Na^+ electrochemical gradient to remove calcium from the cell at the expense of Na^+ entry.

Calcium influx is also via two principal pathways — inflow from the extracellular medium through calcium channels in the surface membrane, and release from internal stores (Fig. 1). The surface membrane calcium channels are of several different types: voltage-controlled channels that open in response to depolarization of the cell membrane, receptor-operated channels that open in response to the binding of an external ligand, second messenger-operated channels that open in response to the binding of a cellular second messenger, and mechanically-operated channels that open in response to mechanical stimulation. A detailed study of voltage-controlled calcium channels is more appropriate in an electrophysiological setting, and so we shall not consider that type of calcium entry here. We also omit the consideration of the other surface membrane channels to concentrate on the properties of calcium release from internal stores.

Calcium release from internal stores such as the ER is the second major calcium influx pathway, and this is mediated principally by two types of calcium channels (but called receptors), the ryanodine receptor, and the inositol (1,4,5)-trisphosphate (IP_3) receptor. The ryanodine receptor, so called because of its sensitivity to the plant alkaloid ryanodine, plays an integral role in excitation-contraction coupling in skeletal and cardiac muscle cells, and is believed to underlie calcium-induced calcium release, whereby a small amount of calcium entering the muscle cell through voltage-gated channels initiates an explosive release of calcium from the sarcoplasmic reticulum. Ryanodine receptors are also found in a variety of non-muscle cells such as neurons, pituitary cells and sea urchin eggs. The IP_3 receptor, although similar in structure to the ryanodine receptor, is found predominantly in non-muscle cells, and is sensitive to the second messenger IP_3.

The exact details of how IP_3 is formed are not particularly relevant to the present discussion. In brief summary, the binding of an extracellular agonist such as a hormone or a neurotransmitter to a receptor in the surface membrane can cause, via a series of reactions that take place in the cell membrane, the production of IP_3 (Fig. 1). The water-soluble IP_3 is free to diffuse through the cell cytoplasm and bind to IP_3 receptors situated on the ER membrane, leading to the opening of these receptors and subsequent release of calcium from the ER.

What makes calcium dynamics so interesting (from a mathematical point of view) are the complicated properties of the IP_3 receptors. Far from being simple release channels that just open and let calcium out, IP_3 receptors are tightly modulated by a host of different factors. Most importantly, the open probability of each receptor is controlled by the local calcium concentration in two different ways. On a fast time scale an increase in local calcium concentration results in an increase in the open probability of the channel, thus leading to positive feedback at fast times. However, on a slower time scale calcium inactivates the IP_3 receptor, restricting the release of further calcium. Hence, a fast positive feedback process is balanced by a slower negative feedback process, in essentially the same manner

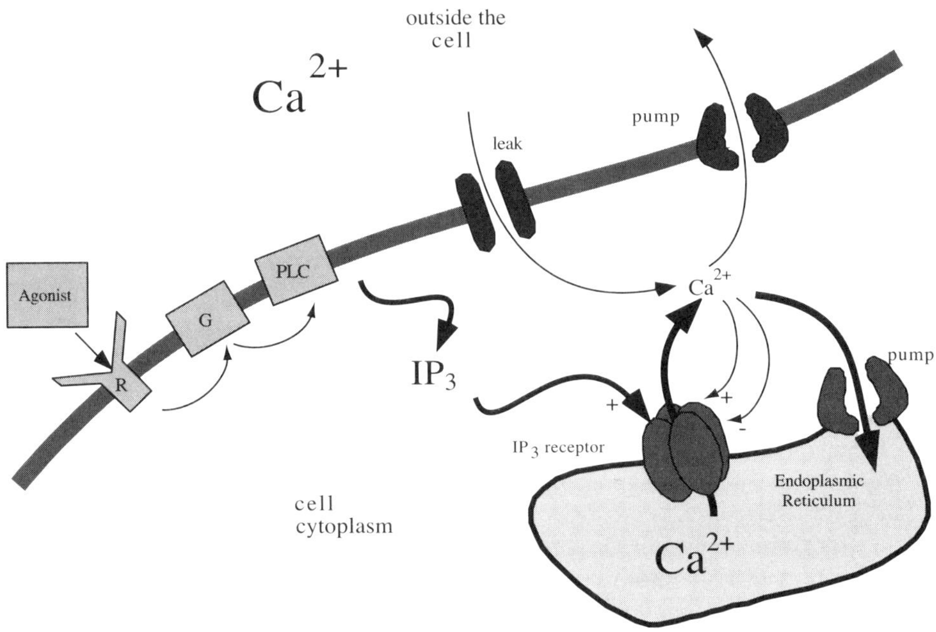

FIGURE 1. Diagram of the pathways involved in the control of cytoplasmic calcium concentration via the production of IP_3.

as we see in the Hodgkin-Huxley and FitzHugh-Nagumo models of membrane excitability. An initial release of calcium through the receptor leads to a fast release of more calcium (resulting in a fast increase in cytosolic calcium concentration). However, once the inactivation kicks in on the slower time scale the IP_3 receptors inactivate, and the calcium concentration is decreased by the action of the various calcium pumps. Depending upon the concentration of IP_3, this sequential positive and negative feedback can lead to cycles of release and uptake of calcium from the ER, and hence calcium oscillations. Furthermore, the excitable nature of calcium release can lead to the appearance of traveling waves of calcium that travel within a single cell, or between multiple cells. Because of the mathematical similarities between models of calcium excitability and models of membrane excitability these dynamic behaviors come as no surprise to the modeler. Reviews of calcium oscillations, and detailed descriptions of the physiology, can be found in Berridge (1993, 1996), Thomas *et al.* (1996), Sanderson *et al.* (1994), and Sneyd *et al.* (1995b).

As an additional control for the cytosolic calcium concentration, calcium is heavily buffered (i.e. bound) by large proteins, with estimates that over 99% of the total cytoplasmic calcium is bound to buffers. This presents additional mathematical complications that we shall not discuss here, but see Wagner and Keizer (1994) and Sneyd *et al.* (1998). The calcium in the internal stores is also heavily buffered.

As a side note, let me describe briefly how calcium oscillations and waves are observed. Essentially, experimentalists fill a cell with a dye that emits light when it binds calcium, i.e., a calcium fluorescent dye. The fluctuations in calcium concentration can then be viewed through a microscope and captured digitally for later analysis. There are, of course, many complications, such as the type of dye used, whether or not it bleaches, whether or not the presence of the dye alters the native cellular behavior, or whether or not the fluorescence can be corrected for cell depth. However, the basic idea is very simple. For a modeler, the important thing to remember is that the waves and oscillations can be easily viewed and analyzed; there is a wealth of real-time data on oscillation shapes and periods, as well as wave properties.

2. Calcium oscillations and waves

Calcium oscillations can be grouped into two major types: those that are dependent on periodic fluctuations of the cell membrane potential and the associated periodic entry of calcium through voltage-gated channels, and those that occur in the absence of extracellular calcium and in the presence of a voltage clamp. Our focus here is on the latter type; the former type are more closely connected with electrical excitability. Within the latter group further distinctions can be made by whether the oscillatory calcium flux is through ryanodine or IP_3 receptors. Here, for the sake of brevity and simplicity, I shall discuss only models of calcium oscillations caused by the generation of IP_3 and the modulation of calcium flux through IP_3 receptors. Nevertheless, as we shall see briefly at the end, ryanodine receptors play an extremely important, though not well understood, role.

Depending on such things as the cell type, or the concentration of IP_3, the period of IP_3-dependent oscillations ranges from a few seconds to a few minutes. There is a great deal of evidence that, in many cell types, these oscillations occur at constant IP_3 concentration and are therefore not driven by oscillations in IP_3 concentration. Although it is risky to generalize, some overall trends can be seen: as the concentration of IP_3 increases, the steady state calcium concentration also increases, the oscillation frequency increases, and the amplitude of oscillation remains approximately constant.

Calcium oscillations usually occur only when the concentration of IP_3 is greater than some critical value, and disappear again when this concentration gets too large. Thus, there is an intermediate range of IP_3 concentrations which generate calcium oscillations.

In many cell types, fluorescent video microscopy has shown that calcium oscillations are organized into repetitive intracellular waves (Rooney and Thomas, 1993), and in large cells such as *Xenopus* oocytes, the intracellular waves develop a high degree of spatial organization, forming concentric circles, plane waves, and multiple spirals (Lechleiter *et al.*, 1991a,1991b; Lechleiter and Clapham, 1992). The speed of intracellular calcium waves is remarkably similar (5–20 μms^{-1}) across a wide range of cell types (Jaffe, 1991). Calcium waves have also been observed propagating from cell to cell across a culture or an organ. These intercellular waves are, in general, independent of extracellular calcium, and can initiate intracellular oscillations in some cell types (Sanderson *et al.*, 1990, 1994; Charles, 1994; Charles *et al.*, 1993, 1996).

Although it is known that calcium controls many cellular processes, the exact significance of calcium oscillations is not completely understood in most cell types. It is widely believed that the oscillations are a frequency-encoded signal, which allow a cell to use calcium as a second messenger while avoiding the toxic effects of prolonged high calcium concentration. A wave enables communication from one side of a cell to the other, and serves to synchronize an overall cellular response

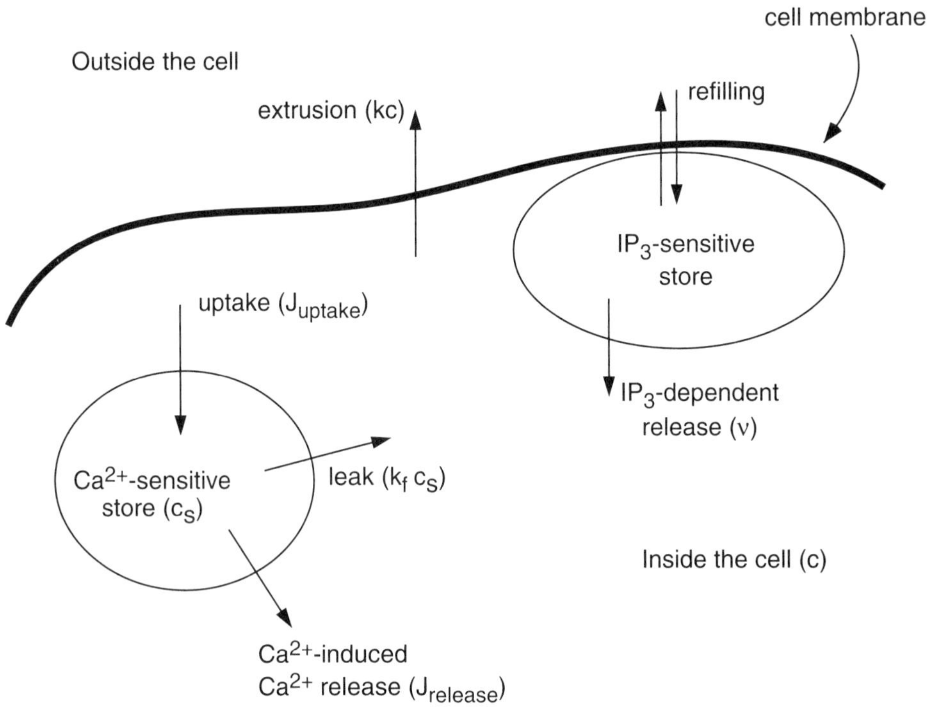

FIGURE 2. Schematic diagram of the two-pool model of calcium oscillations.

to a local stimulus. In a similar fashion, an intercellular calcium wave serves to coordinate a multicellular response.

3. The two-pool model

One of the earliest models for IP_3-dependent calcium release assumes the existence of two distinct internal calcium stores, one of which is sensitive to IP_3, the other of which is sensitive to calcium (Kuba and Takeshita, 1981; Goldbeter *et al.*, 1990; Goldbeter, 1996). Agonist stimulation leads to the production of IP_3 which releases calcium from the IP_3-sensitive store through IP_3 receptors. The calcium that is thereby released stimulates the release of further calcium from the calcium-sensitive store, possibly via ryanodine receptors. A crucial assumption of the model is that the concentration of calcium in the IP_3-sensitive store remains constant, as the store is quickly refilled from the extracellular medium. A schematic diagram of the model is given in Fig. 2.

Recent work by Dupont and Goldbeter (1993, 1994) has shown that the model does not depend on the existence of two separate pools of calcium; the model equations can equally well be used to describe the release of calcium from a single pool, with the release modulated by both IP_3 and calcium. Nevertheless, for convenience, we persist in calling this the two-pool model.

Let c denote the concentration of calcium in the cytoplasm, and c_s denote the concentration of calcium in the calcium-sensitive pool. We assume that IP_3 causes a steady flux, ν, of calcium into the cytosol, and that calcium is pumped out of the

k	=	$10\ \text{s}^{-1}$	K_1	=	$1\ \mu\text{M}$
K_2	=	$2\ \mu\text{M}$	K_3	=	$0.9\ \mu\text{M}$
V_1	=	$65\ \mu\text{Ms}^{-1}$	V_2	=	$500\ \mu\text{Ms}^{-1}$
k_f	=	$1\ \text{s}^{-1}$	m	=	2
n	=	2	p	=	4

TABLE 1. Typical parameter values for the two-pool model for calcium oscillations (Goldbeter *et al.*, 1990).

cytoplasm at the rate $-kc$. Then

$$\frac{dc}{d\tau} = \nu - kc - \tilde{f}(c, c_s), \tag{1}$$

$$\frac{dc_s}{d\tau} = \tilde{f}(c, c_s), \tag{2}$$

$$\tilde{f}(c, c_s) = J_{\text{uptake}} - J_{\text{release}} - k_f c_s, \tag{3}$$

where

$$J_{\text{uptake}} = \frac{V_1 c^n}{K_1^n + c^n}, \tag{4}$$

$$J_{\text{release}} = \left(\frac{V_2 c_s^m}{K_2^m + c_s^m}\right)\left(\frac{c^p}{K_3^p + c^p}\right), \tag{5}$$

and τ denotes time. The flux of calcium from the cytoplasm into the calcium-sensitive pool is given by $\tilde{f}$; J_{uptake} is the rate at which calcium is pumped from the cytosol into the calcium-sensitive pool by an active process, and J_{release} is the rate at which calcium is released from the calcium-sensitive pool. Note that, as c increases, so does J_{release}. Thus, calcium stimulates its own release through positive feedback, usually called calcium-induced calcium release, or CICR (Endo *et al.*, 1970; Fabiato, 1983). It is this positive feedback that is central to the model's behavior. Finally, the rate at which calcium leaks from the calcium sensitive pool into the cytosol is $k_f c_s$. In the model, ν is constant for constant IP_3 concentration, and is treated as a control parameter. Thus, the behavior of the model at different constant IP_3 concentrations can be studied by varying ν.

For convenience we nondimensionalize the model equations. Let $u = c/K_1$, $t = \tau k$, $v = c_s/K_2$, $\alpha = K_3/K_1$, $\beta = V_1/V_2$, $\gamma = K_2/K_1$, $\delta = k_f K_2/V_2$, $\mu = \nu/(kK_1)$ and $\epsilon = kK_2/V_2$, to find

$$\frac{du}{dt} = \mu - u - \frac{\gamma}{\epsilon} f(u, v), \tag{6}$$

$$\frac{dv}{dt} = \frac{1}{\epsilon} f(u, v), \tag{7}$$

$$f(u, v) = \beta\left(\frac{u^n}{u^n + 1}\right) - \left(\frac{v^m}{v^m + 1}\right)\left(\frac{u^p}{\alpha^p + u^p}\right) - \delta v. \tag{8}$$

If the exchange of calcium between the cytosol and the calcium-sensitive pool is fast (i.e., V_1 and V_2 are large), then ϵ is a small parameter. A table of typical parameter values in the model is given in Table 1. For these values, $\epsilon \approx 0.04$.

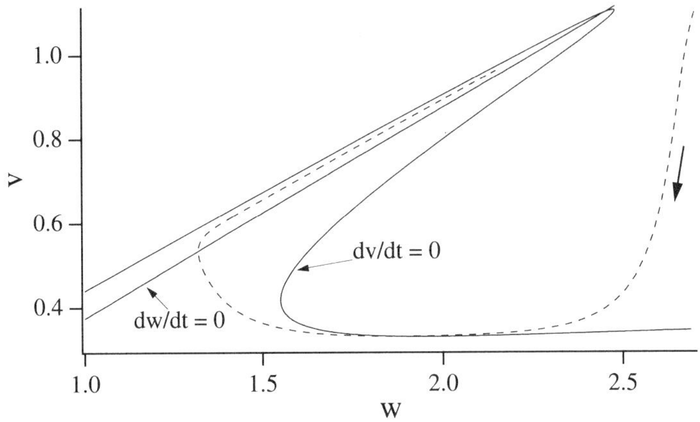

FIGURE 3. Nullclines and sample trajectory of the two-pool model transformed to w, v coordinates (Sneyd *et al.*, 1993; Fig. 4).

3.1. Excitability and Oscillations in the two-pool model. The two-pool model can be put into the form of a generalized FitzHugh-Nagumo model (Keener and Sneyd, 1998) by a simple change of variables. If we let $w = u + \gamma v$ then the two-pool model becomes

$$\frac{dw}{dt} = \mu - (w - \gamma v), \tag{9}$$

$$\frac{dv}{dt} = \frac{1}{\epsilon} f(w - \gamma v, v) = \frac{1}{\epsilon} F(w, v). \tag{10}$$

The nullclines of the transformed equations are shown in Fig. 3, where it can be seen that one nullcline is N-shaped, while the other is a straight line, as in the FHN model. Thus, the analysis of the FitzHugh-Nagumo model can be applied, essentially without change, to the temporal behavior of the two-pool model. It is therefore not surprising that the two-pool model is excitable and exhibits oscillatory behavior. When μ is slightly below the lower Hopf bifurcation point (calculated below), a subthreshold addition of calcium gives a small response, while a superthreshold addition causes a large transient before the return to steady state.

The steady state of the two-pool model, (u_0, v_0), is given by

$$u_0 = \mu, \tag{11}$$

$$f(\mu, v_0) = 0, \tag{12}$$

and the stability of the steady state is determined by the roots of the characteristic equation

$$\lambda^2 + H\lambda - \frac{f_v}{\epsilon} = 0 \tag{13}$$

where

$$H = \frac{\gamma f_u(u_0, v_0)}{\epsilon} - \frac{f_v(u_0, v_0)}{\epsilon} + 1. \tag{14}$$

Since $f_v < 0$, the roots of (13) have negative real part (and thus the steady state is stable) if $H > 0$, and have positive real part if $H < 0$. At $H = 0$ the steady

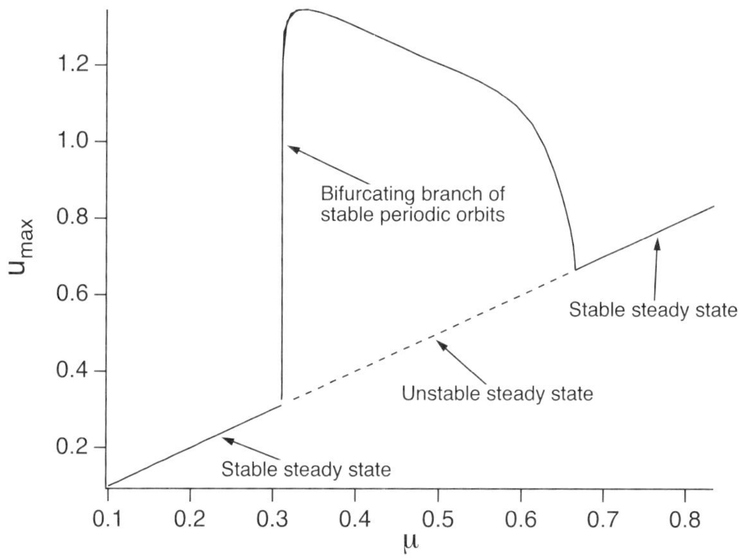

FIGURE 4. Bifurcation diagram of the two-pool model, using the parameter values given in Table 1. As μ is increased, oscillations appear at a supercritical Hopf bifurcation, and disappear in the same manner. Hence, the model is oscillatory only for intermediate values of μ, which corresponds to the concentration of IP_3. (Sneyd *et al.*, 1993, Fig. 2b.)

state changes stability through a Hopf bifurcation, and at these points a branch of periodic orbits appears. The amplitude and period of these periodic orbits as a function of the bifurcation parameter μ can be tracked with the use of the software package AUTO (Doedel, 1986), and the results are shown in Fig. 4.

Both Hopf bifurcations are supercritical and the two bifurcation points are connected by a branch of stable periodic orbits. Typical oscillations are shown in Fig. 5. Note that oscillations occur for a constant value of μ, i.e., constant IP_3 concentration. Thus, the two-pool model shows that calcium-induced calcium release is sufficient to produce oscillations in the absence of IP_3 oscillations. The function of IP_3 is to produce a steady influx of calcium into the cytosol from the IP_3-sensitive pool, and this steady influx drives the oscillations.

3.2. Traveling waves in the two-pool model. In many cells, calcium oscillations are organized into repetitive intracellular waves. Although there is controversy about the exact mechanisms by which these waves propagate (and it is certainly true that the mechanisms differ from cell type to cell type), it is widely believed that, in many cell types, calcium waves are driven by the diffusion of calcium between release sites. In this hypothesis, the calcium released from one pool diffuses to neighboring pools and initiates CICR there. Repetition of this process can then generate an advancing wave front of high calcium concentration, i.e., a calcium wave. That this is a possible mechanism for wave propagation is confirmed by the two-pool model. Inclusion of calcium diffusion into the (nondimensional)

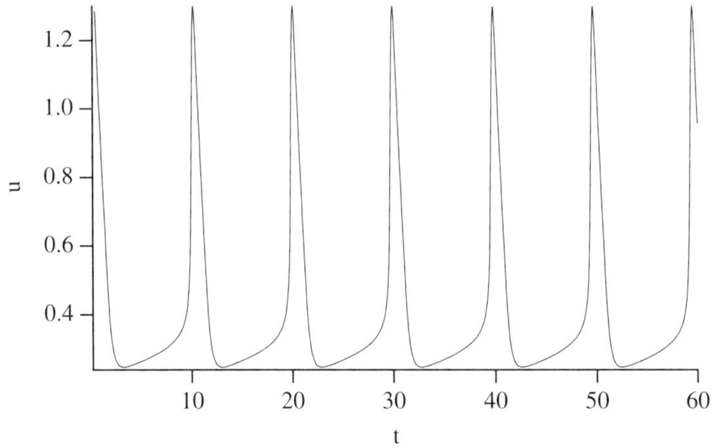

FIGURE 5. Oscillations of the two-pool model when $\mu = 0.4$. (Sneyd *et al.*, 1993, Fig. 3.)

model equations gives

$$\epsilon u_t = \epsilon^2 u_{xx} + \epsilon(\mu - u) - \gamma f(u, v), \tag{15}$$

$$\epsilon v_t = f(u, v), \tag{16}$$

where the spatial variable, x, has been scaled by $\epsilon/\sqrt{D}$, D being the diffusion coefficient of calcium. Although there is some difficulty in defining a diffusion coefficient of calcium in the cytosol (because calcium is so extensively buffered, and may not obey a simple diffusion equation), we ignore these complications. Diffusion of v is not included in (16) since we assume that calcium in the store is not free to diffuse.

Although the temporal behavior of the model can be understood by comparison with the FitzHugh-Nagumo model, the same is not true of the model when calcium diffusion is included. In particular, when diffusion is included equations (15) and (16) cannot be put into the same form as the FitzHugh-Nagumo model, and therefore previous theories of wave propagation need not apply. An example of this is discussed in Keener and Sneyd (1998; Chapter 12).

The two-pool model is capable of reproducing the waves observed in a variety of cell types (Dupont and Goldbeter, 1994). By varying the parameter values, the model can produce a wave with a sharp front and a sharp back, as is observed in cardiac cells, as well as the smoother rise followed by a homogeneous return to steady levels, as is observed in hepatocytes. Furthermore, in two spatial dimensions the model can exhibit spiral waves similar to those observed in *Xenopus* oocytes (Girard *et al.*, 1992). This model is still used today by a number of modelling groups, and is certainly the model that is most widely known among experimentalists. Nevertheless, it has some deficiencies, as we discuss below, and has been largely superseded by more recent models.

4. Mechanistic models for calcium release

Although the two-pool model reproduces experimental data extremely well, both qualitatively and quantitatively, recent experimental evidence indicates that the role of calcium is more complicated than was assumed in this model. In the two-pool model calcium stimulates its own release (thus the term $c^p/(K_3^p + c^p)$ in (5)), while the flow of calcium from the internal store is terminated when the concentration of calcium in the internal store becomes too low (thus the term $c_s^m/(K_2^m + c_s^m)$). However, it is now known that not only does calcium stimulate its own release, it also inhibits it, but on a slower time scale (Parker and Ivorra, 1990; Finch *et al.*, 1991; Bezprozvanny *et al.*, 1991). It is hypothesized that this sequential activation and inactivation of the IP_3 receptor by calcium (rather than the depletion of the internal store) is the fundamental mechanism underlying IP_3-dependent calcium oscillations and waves, and a number of models incorporating this hypothesis have appeared (Sneyd *et al.*, 1995b).

4.1. A detailed receptor model. One approach to determining whether sequential activation and inactivation of the IP_3 receptor by calcium can produce calcium oscillations is to construct a detailed model of the IP_3 receptor, including all the possible receptor states and transitions between them (De Young and Keizer, 1992). To do so, we assume that the IP_3 receptor consists of three equivalent and independent subunits, all of which must be in a conducting state before the receptor allows calcium flux. Each subunit has an IP_3 binding site, an activating calcium binding site, and an inactivating calcium binding site, each of which can be either occupied or unoccupied, and thus each subunit can be in one of eight states. Each state of the subunit is labelled S_{ijk}, where i, j and k are equal to 0 or 1, with a 0 indicating the binding site is unoccupied and a 1 indicating it is occupied. The first index refers to the IP_3 binding site, the second to the calcium activation site, and the third to the calcium inactivation site. This is illustrated in Fig. 6. Although a fully general model would include 24 rate constants, we make a number of simplifying assumptions. For instance, the rate constants are assumed to be independent of whether activating calcium is bound or not. Further, the kinetics of calcium activation are assumed to be independent of whether IP_3 is bound or not, and finally, calcium inactivation is assumed to be independent of calcium activation and of IP_3. This leaves only 10 independent rate constants, $k_1, \ldots, k_5$, and $k_{-1}, \ldots, k_{-5}$.

The fraction of subunits in the state S_{ijk} is denoted by x_{ijk}. The differential equations for these are based on mass action kinetics, and thus, for example,

$$\frac{dx_{000}}{dt} = -(V_1 + V_2 + V_3), \tag{17}$$

where

$$V_1 = k_1 p x_{000} - k_{-1} x_{100}, \tag{18}$$

$$V_2 = k_4 c x_{000} - k_{-4} x_{001}, \tag{19}$$

$$V_3 = k_5 c x_{000} - k_{-5} x_{010}, \tag{20}$$

where p denotes IP_3 concentration and c denotes calcium concentration. V_1 describes the rate at which IP_3 binds to and leaves the IP_3 binding site, V_2 describes the rate at which calcium binds to and leaves the inactivating site, and similarly for V_3. Since experimental data indicate that the receptor subunits act in a cooperative

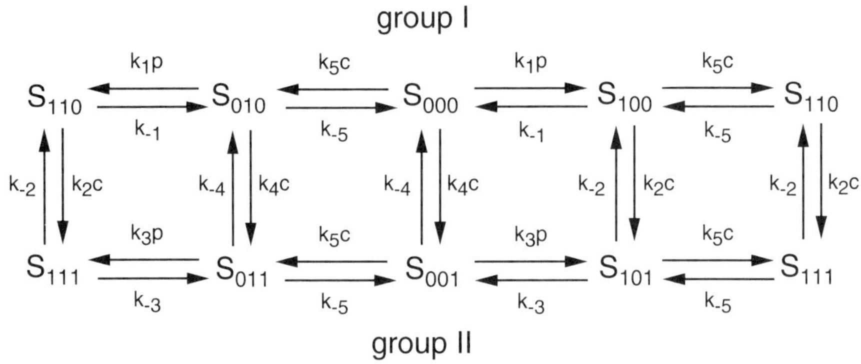

FIGURE 6. The binding diagram for the receptor model. The first subscript refers to the IP_3 binding site, the second to the calcium activation site, and the third to the calcium inactivation site; a 1 denotes an occupied binding site, a 0 denotes an empty binding site. Here, c denotes calcium concentration, and p denotes IP_3 concentration.

fashion, the model assumes that the IP_3 receptor passes calcium current only when three subunits are in the state S_{110} (i.e., with one IP_3 and one activating calcium bound), and thus the open probability of the receptor is x_{110}^3.

The set of seven differential equations for the receptor states (there are eight receptor states, but only seven are independent, as the x_{ijk}s must sum to one) are combined with a differential equation for calcium transport to obtain the full model

$$\frac{dc}{dt} = \overbrace{(\nu_1 x_{110}^3 + \nu_2)(c_s - c)}^{\text{receptor flux}} - \overbrace{\frac{\nu_3 c^2}{c^2 + k_3^2}}^{\text{pumping}}, \tag{21}$$

where c denotes the concentration of calcium in the cytoplasm, and c_s denotes the concentration in the ER. ν_1 and ν_2 are constants. The first term is the calcium flux through the IP_3 receptor, and is proportional to the concentration difference between the ER and the cytoplasm. It includes an IP_3-independent leak (ν_2) from the ER into the cytoplasm. The second term, similar to the pump term in the two-pool model, describes the action of the calcium ATPases that pump calcium from the cytoplasm into the ER, and is based on experimental data that shows that the calcium ATPase is cooperative, with a Hill coefficient of 2. For simplicity, we assume that the cell is closed, i.e., that there is no calcium exchange between the inside and outside of the cell. In this case, c_s is determined by constraining the total amount of intracellular calcium to be constant, and thus

$$c_{\text{tot}} = v_c c_s + c, \tag{22}$$

where v_c is the ratio of the ER volume to the cytoplasmic volume. However, this somewhat artificial constraint can be removed without affecting model behavior. For if calcium exchange between the inside and the outside of the cell is much slower than exchange between the cytoplasm and the ER, we can apply a quasi-steady state hypothesis and assume that the calcium concentration is constant on a fast time scale.

k_1	=	$400\ \mu\text{M}^{-1}\text{s}^{-1}$	k_{-1}	=	$52\ \text{s}^{-1}$
k_2	=	$0.2\ \mu\text{M}^{-1}\text{s}^{-1}$	k_{-2}	=	$0.21\ \text{s}^{-1}$
k_3	=	$400\ \mu\text{M}^{-1}\text{s}^{-1}$	k_{-3}	=	$377.2\ \text{s}^{-1}$
k_4	=	$0.2\ \mu\text{M}^{-1}\text{s}^{-1}$	k_{-4}	=	$0.029\ \text{s}^{-1}$
k_5	=	$20\ \mu\text{M}^{-1}\text{s}^{-1}$	k_{-5}	=	$1.64\ \text{s}^{-1}$
c_{tot}	=	$2\ \mu\text{M}$	ν_1	=	$1.11\ \text{s}^{-1}$
v_c	=	0.185	ν_2	=	$0.02\ \text{s}^{-1}$
k_3	=	$0.1\ \mu\text{M}$	ν_3	=	$0.9\ \mu\text{M}^{-1}\text{s}^{-1}$

TABLE 2. Parameters of the receptor model (De Young and Keizer, 1992) for calcium oscillations.

Parameters in the model were chosen to obtain agreement with the steady state open probability of the receptor, as measure by Bezprozvanny *et al.* (1991). However, the kinetic properties of the IP_3 receptor are equally important; the receptor is activated quickly by calcium, but inactivated by calcium on a slower time scale. In the model, this is incorporated in the magnitude of the rate constants, and is the basis of a simplification of the model that we discuss below.

As the concentration of IP_3 is increased, periodic orbits appear via a supercritical Hopf bifurcation, and disappear in the same manner. For many parameters in the physiological range, the two Hopf bifurcations are connected by a branch of periodic orbits and the period of the orbits is a decreasing function of IP_3 concentration, as observed experimentally. This behavior, similar to that seen in the two-pool model, seems to be typical of many models of calcium oscillations.

4.2. Reduction of the detailed receptor model. The complexity of the receptor model (eight differential equations and numerous parameters) provides ample motivation to seek a simpler model that retains its essential properties. Since IP_3 binds quickly to its binding site, and calcium binds quickly to the activating site, we can dispense with the transient details of these binding processes, and assume instead that the receptor is in quasi-steady state with IP_3 binding and calcium activation (De Young and Keizer, 1992; Keizer and De Young, 1994; Li and Rinzel, 1994; Tang *et al.*, 1996).

As shown in Fig. 6, the receptor states are arranged into two groups: those without calcium bound to the inactivating site ($S_{000}, S_{010}, S_{100}$ and S_{110}, shown in the upper line of Fig. 6; called group I states), and those with calcium bound to the inactivating site ($S_{001}, S_{011}, S_{101}$ and S_{111}, shown in the lower line of Fig. 6; called group II states). Because the binding of IP_3, and the binding of calcium to the activating site are assumed to be fast processes, it follows that, within each group, the binding states are at quasi-steady state with respect to transitions within the group. More precisely, write the differential equations governing the transitions between the states in group I, say, to get

$$\frac{dx_{000}}{dt} = -x_{000}(k_5c + k_1p + k_4c) + k_{-5}x_{010} + k_{-1}x_{100} + k_{-4}x_{001}, \tag{23}$$

$$\frac{dx_{100}}{dt} = -x_{100}(k_5c + k_{-1} + k_2c) + k_{-5}x_{110} + k_1px_{000} + k_{-2}x_{101}, \tag{24}$$

$$\frac{dx_{010}}{dt} = -x_{010}(k_{-5} + k_1p + k_4c) + k_5cx_{000} + k_{-1}x_{110} + k_{-4}x_{011}. \tag{25}$$

where c denotes calcium concentration and p denotes IP_3 concentration. The differential equation for the fourth receptor state, x_{110} is superfluous, as we have the constraint

$$x_{000} + x_{010} + x_{100} + x_{110} = 1 - y, \tag{26}$$

where

$$y = x_{001} + x_{011} + x_{101} + x_{111}. \tag{27}$$

Since these four states are assumed to be in quasi-steady state, we set $dx_{000}/dt = dx_{100}/dt = dx_{010}/dt = 0$, and solve for the group I state probabilities. This gives, for example,

$$x_{000} = \frac{K_1 K_5 (1-y)}{(p+K_1)(c+K_5)}, \tag{28}$$

where $K_i = k_{-i}/k_i$. An identical procedure applied to the group II receptor states gives the quasi-steady state equations for the group II states.

It now remains to derive a differential equation for y. Note that y changes only on a slow time scale since any changes in y involve calcium leaving or binding to the inactivating site, a process which is assumed to be slow. If we write down the differential equations for the group II sites, taking care to include the transitions between the group I and group II sites, add the four equations, and substitute in the quasi-steady state expressions we get, finally,

$$\frac{dy}{dt} = \left[\frac{(k_{-2}K_2K_1 + k_{-2}pK_4)c}{K_4K_2(p+K_1)}\right](1-y) - \left(\frac{k_{-4}p + k_{-2}K_3}{p+K_3}\right)y. \tag{29}$$

This can easily be written in the form

$$\tau_y(c,p)\frac{dy}{dt} = y_\infty(c,p) - y, \tag{30}$$

which is useful for comparison with other models.

It is now a relatively simple matter to show how the reduced receptor model can be used to construct a simpler model for calcium oscillations. First, recall that the equation governing the calcium dynamics is (21)

$$\frac{dc}{dt} = (\nu_1 x_{110}^3 + \nu_2)(c_s - c) - \frac{\nu_3 c^2}{c^2 + k_3^2}. \tag{31}$$

Into this equation we substitute the expression for x_{110},

$$x_{110} = \frac{pc(1-y)}{(p+K_1)(c+K_5)}, \tag{32}$$

and the differential equation for y, to get a model for calcium oscillations consisting of two differential equations rather than the original eight.

Note that $1-y$, which is the proportion of receptors that are not inactivated by calcium, plays the role of an inactivation variable, similar in spirit to the variable h in the Hodgkin-Huxley equations. To emphasize this similarity, the reduced model can be written in the form

$$x_{110} = \frac{pc}{(p+d_1)(c+d_5)}h, \tag{33}$$

$$\tau_h(c,p)\frac{dh}{dt} = h_\infty(c,p) - h, \tag{34}$$

where $h = 1 - y$, and τ_h and h_∞ are easily calculated from the corresponding differential equation for y.

4.3. A simpler model. An alternative approach to modelling calcium release is based on heuristic assumptions about the kinetic properties of the receptor (Atri *et al.*, 1993), and is similar in spirit to the construction of the Hodgkin-Huxley model from voltage-clamp data. We assume first that the IP_3 receptor consists of three binding domains, the first of which binds IP_3, the other two binding calcium, and second, that the receptor passes calcium current only when IP_3 is bound to domain 1, calcium is bound to domain 2, but calcium is *not* bound to domain 3. Thus, calcium activates the receptor by binding to domain 2, and inactivates the receptor by binding to domain 3. Each binding domain consists of a number of binding sites, grouped on the basis of functionality. If p_1 is the probability that IP_3 is bound to domain 1, p_2 is the probability that calcium is bound to domain 2, and $1 - p_3$ is the probability that calcium is bound to domain 3, then, assuming independence of the domains, it follows that the steady state calcium flux through the IP_3 receptor, J_{channel}, is given by

$$J_{\text{channel}} = k_f p_1 p_2 p_3, \tag{35}$$

for some constant k_f. The probabilities $p_i, i = 1, 2, 3$ are chosen so that J_{channel} agrees with the steady state experimental data of Parys *et al.* (1992) from *Xenopus* oocytes. Good agreement with data is obtained by choosing

$$p_1 = \mu_0 + \frac{\mu_1 p}{p + k_\mu}, \tag{36}$$

$$p_2 = b + \frac{(1-b)c}{k_1 + c}, \tag{37}$$

$$p_3 = \frac{k_2^2}{k_2^2 + c^2}, \tag{38}$$

where p denotes IP_3 concentration, c denotes the calcium concentration, and where μ_0, μ_1, b, k_1 and k_2 are constants.

To complete the model, it is assumed that p_1 and p_2 are instantaneous functions of the calcium and IP_3 concentrations, but that p_3 acts on a slower time scale, so that

$$J_{\text{channel}} = k_f p_1 p_2 h, \tag{39}$$

where h is a time-dependent inactivation variable, satisfying the differential equation

$$\tau_h \frac{dh}{dt} = \frac{k_2^2}{k_2^2 + c^2} - h. \tag{40}$$

Thus,

$$\frac{dc}{dt} = \overbrace{k_f \left(\mu_0 + \frac{\mu_1 p}{p + k_\mu}\right)\left(b + \frac{(1-b)c}{k_1 + c}\right) h}^{\text{receptor flux}} - \overbrace{\frac{\gamma c}{k_\gamma + c}}^{\text{pumping}} + \beta. \tag{41}$$

In a fashion similar to the De Young-Keizer model, the term $\gamma c/(k_\gamma + c)$ represents pumping of calcium out of the cytoplasm, and β represents a constant leak into the cytoplasm.

It is important to note some features of this heuristic model. First, the model equations are of the same form as the reduced receptor model discussed above, and of the same form as the Hodgkin-Huxley and FitzHugh-Nagumo models. Obviously, this combination of fast variables and slow variables, with inactivation acting on a slow time scale, is a feature common to many physiological systems. The variable h in the heuristic model has a simpler interpretation than does the variable h in the reduced receptor model, and the functions τ_h and h_∞ have a simpler form. However, in attaining greater simplicity, some accuracy has been lost. For instance, in the heuristic model τ_h is a constant, independent of c and p; the more detailed receptor model, and its reduced version, find τ_h to be a function of c and p, i.e., the kinetics of receptor inactivation are affected by the concentrations of calcium and IP_3.

The heuristic model exhibits oscillations in a manner similar to the models discussed above. For a wide range of parameter values, two Hopf bifurcation points exist and are connected by a branch of stable periodic orbits. This is behavior typical of models of calcium oscillations.

A number of other points about the heuristic model are worth noting. First, it does not include the factor $c_s - c$ in the term describing the IP_3-sensitive calcium current. Thus, it assumes that the concentration of calcium in the ER is so high, and so well buffered, that depletion of the ER has only a negligible effect on intracellular calcium dynamics for most of the physiological regime. Because of this omission, the structure of the heuristic model is different from the usual FitzHugh-Nagumo system, and the model becomes unphysiological when τ_h is too large. Nevertheless, in the physiological regime the heuristic model agrees well with experimental data. Also, the form of the pumping term is different from that in the detailed receptor model, which uses a Hill function with coefficient 2 (Keener and Sneyd, 1998, Chapter 1). There is experimental evidence that the form used in the receptor model is a more accurate description of the calcium ATPase found in a variety of cell types but it is not clear how this change in the pumping term affects model behavior. The differences and similarities between the detailed receptor model and the heuristic model underline the fact that there are many choices to make in the construction of even the simplest model. The obvious question to ask is to what extent the choice of assumptions affects the final results. In other words, how sensitive are the model predictions to the underlying assumptions? How much complication is vital, and how much is a waste of time and money? In general these are difficult questions to answer, and can be answered completely only after detailed (and time-consuming) comparisons between the models. For this reason such comparisons are rarely performed, or at least not before experimental data indicates that the comparison will lead to useful distinctions. Despite these difficulties, however, the extensive similarities between these two models (models that were obtained independently by different methods) suggest strongly that fast activation and slow inactivation of the IP_3 receptor by calcium are the crucial underlying mechanisms that generate calcium oscillations.

4.4. Calcium waves in Xenopus. Although so far we have concentrated on models for intracellular calcium oscillations, one important goal of such modelling work is to understand and explain the mechanisms underlying the propagating calcium waves observed in many cell types, particularly *Xenopus* oocytes. In 1991, it was discovered by Lechleiter and Clapham and their coworkers that intracellular calcium waves in immature *Xenopus* oocytes showed remarkable spatio-temporal

b	=	0.11	k_f	=	8.1 μMs^{-1}
β	=	0.02 μMs^{-1}	D_p	=	300 μm^{2}s^{-1}
γ	=	2 μMs^{-1}	D_c	=	20 μm^{2}s^{-1}
τ_h	=	2 s	k_p	=	$0 - 0.2$ s^{-1}
k_1	=	0.7 μM	k_μ	=	4 μM
k_2	=	0.7 μM	μ_0	=	0.567
k_γ	=	0.1 μM	μ_1	=	0.433

TABLE 3. Standard parameters of the model for calcium waves. One important point to note is the difference between D_p and D_c. Although IP_3 is a much larger molecule than calcium, it diffuses faster through the cell because it is not buffered. Fast calcium buffering can greatly decrease the effective diffusion coefficient of calcium (Wagner and Keizer, 1994).

organization. By loading the oocytes with a calcium-sensitive dye, releasing IP_3, and observing calcium release patterns with a confocal microscope, Lechleiter and Clapham (1992; Lechleiter *et al.*, 1991) observed concentric and spiral calcium waves *in vivo*. The crucial feature of *Xenopus* oocytes that makes these observations possible is their large size. *Xenopus* oocytes can have a diameter as large as 1000 μm, almost two orders of magnitude greater than most other cells. In a small cell, a typical calcium wave (often with a width of close to 100 μm) cannot be observed in its entirety, and there is not enough room for a spiral or a complete concentric wave to form. However, in a large cell it may be possible to observe both the wave front and the wave back, as well as spiral waves, and this has made the *Xenopus* oocyte one of the most important systems for the study of calcium waves. Of course, what is true for *Xenopus* oocytes is not necessarily true for other cells, and so one must be cautious about extrapolating to other cell types.

To model calcium waves in *Xenopus* oocytes, we choose to use the heuristic model, although any of the other models would give similar results. We extend (41) and (40) by the addition of calcium diffusion, and the diffusion and degradation of IP_3. Thus,

$$\frac{\partial p}{\partial t} = D_p\nabla^2 p - k_p p. \tag{42}$$

$$\frac{\partial c}{\partial t} = D_c\nabla^2 c + k_f\left(\mu_0 + \frac{\mu_1 p}{p+k_\mu}\right)\left(b + \frac{(1-b)c}{k_1+c}\right)h - \frac{\gamma c}{k_\gamma + c} + \beta, \tag{43}$$

$$\tau_h\frac{dh}{dt} = \frac{k_2^2}{k_2^2+c^2} - h. \tag{44}$$

D_p and D_c are the diffusion coefficients of IP_3 and calcium respectively.

Equations (42)–(44) were solved on a square domain with no-flux boundary conditions. Standard parameters of the calcium wave model are given in Table 3. We simulated the release of a bolus of IP_3 into the cell (a common experimental protocol). As the IP_3 diffuses out from the point of release it forms either single or multiple wave fronts. If enough IP_3 is injected to make the cytoplasm oscillatory then periodic waves are formed. (We expect this from the standard theory; see, for example, Kopell and Howard, 1973; Duffy *et al.*, 1980; Maginu, 1985; Neu, 1979; Murray, 1989). If the cytoplasm is excitableexcitable!cytoplasm, but not oscillatory

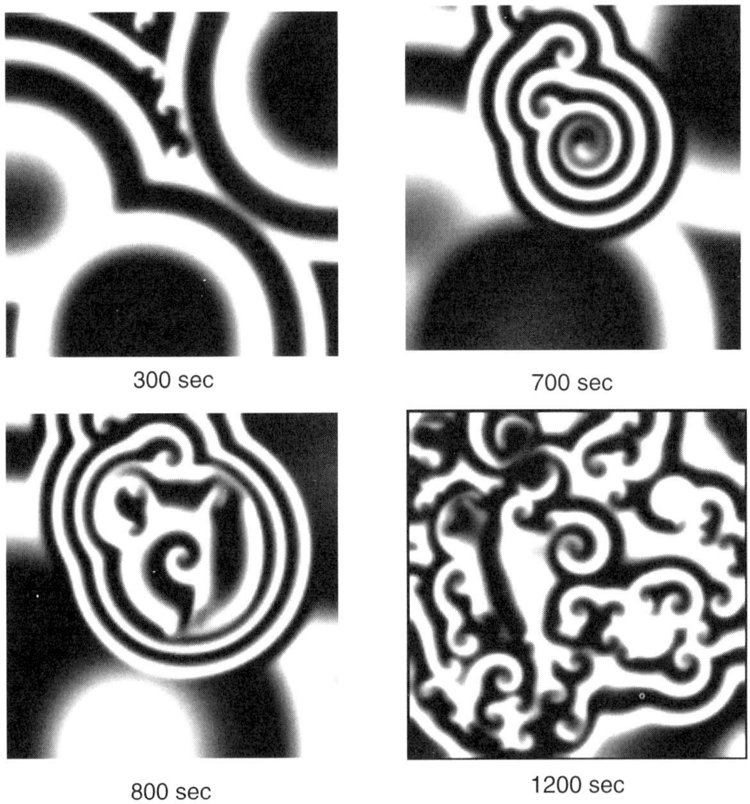

FIGURE 7. Pictures of spiral breakup and periodic reforming of spirals. Full details of these computations are given in McKenzie and Sneyd, 1998.

(i.e., if slightly less IP_3 is injected) then a single wave will form. When a wave front is broken, a spiral wave of calcium often forms. Depending on the initial conditions, these spiral waves can be stable or unstable. In the unstable case, the branches of the spiral can intersect themselves, and cause breakup of the spiral (Fig. 7).

5. Two case studies of experimental/theoretical collaboration

When doing mathematical physiology one is regularly challenged by both mathematicians and physiologists. A common comment from the mathematical literati is "Where's the math?", while from the experimental side one often hears "But how can all this math stuff help me? It's really just a waste of time". Because of the prevalence of these attitudes it is important to address them both. Of course, what follows here is a purely personal point of view, not necessarily guaranteed to coincide with views of most, or even any, other modelers.

The response to the first question is simple. The whole point of mathematical physiology is to use mathematical techniques to solve an important physiological problem. Nowhere is it specified that one must use complex, deep and nontrivial mathematics; the requirement is only that an important physiological problem is addressed. Those who ask where the difficult math is are missing the fundamental point of modelling.

However, the second question is much more difficult. How can one use models to answer questions that, firstly, physiologists care about, and, secondly, that they can't answer for themselves? There is, of course, no ready answer to this. I shall answer merely by describing two examples where a model has been able to propose, or answer, questions important to the experimentalists. For convenience these example are taken from my own work; there are many other examples I could have chosen.

5.1. Calcium oscillations in pancreatic acinar cells. Pancreatic acinar cells are the majority cell type in the pancreas, and their function is the production and secretion of digestive enzymes. The pancreatic acinus is a particularly interesting system in which to study calcium waves (Nathanson *et al.*, 1992; Kasai, 1995; Yule *et al.*, 1996; Pfeiffer *et al.*, 1998). Firstly, the pancreatic acinar cell is electrically non-excitable, and the calcium wave results (at least in large part) from the release of calcium from the endoplasmic reticulum. Secondly, each acinus consists of a number of cells arranged in a ring around a central duct, and the calcium wave travels from cell to cell around this ring in a characteristic fashion (Yule *et al.*, 1996). The function of this intercellular calcium wave is not entirely clear, although it appears to increase the efficiency of enzyme secretion of the acinus, presumably by coordinating the secretion of each individual cell with that of its neighbors. Thirdly, and most importantly for this example, different agonists cause different wave responses (Lawrie *et al.*, 1993; Thorn *et al.*, 1993). This, in turn, results from the fact that different agonists cause different oscillatory responses in each individual cell. Application of intermediate doses of acetylcholine (ACh) or carbachol (CCh) causes high-frequency oscillations superimposed on a raised baseline, while application of cholecystokinin (CCK) causes baseline oscillations with a much larger period (Yule *et al.*, 1991).

Despite these differences, there is much evidence that both types of oscillations result from an agonist-dependent increase in intracellular IP_3 concentration. How might these two very different patterns of calcium release occur, with apparently the same basic intracellular processes? In particular, how do the long baseline spikes observed with CCK occur, given that the IP_3 concentration is thought to be maintained above basal levels throughout the period of stimulation, whereas the calcium concentration, which initially provides the negative feedback signal to end calcium flux from the stores, has returned to its basal values long before the next spike occurs? Our goal was to understand the mechanisms underlying these different oscillatory patterns, and provide a single explanation for these seemingly diverse phenomena.

Our model (LeBeau *et al.*, 1999) began with the hypothesis that different agonists caused different modulations of the IP_3 receptor. For physiological reasons, the most plausible hypothesis was that CCK caused the IP_3 receptor to be phosphorylated, and thus inactivated, while ACh did not. First we constructed a model to check that such a hypothesis was a possible explanation for the observed differences between the oscillatory patterns.

5.1.1. *Model construction.* We assume that the complete IP_3 receptor is composed of four, functionally identical, independent subunits. The reaction scheme governing transitions of each subunit is shown in Figure 8. S denotes the fraction

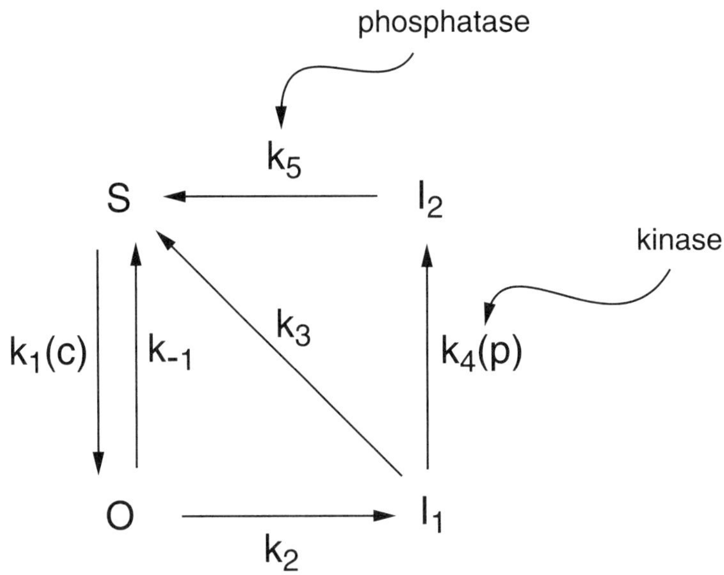

FIGURE 8. Diagram of the IP_3 receptor model used to model calcium oscillations in pancreatic acinar cells.

of subunits in the shut state (S), in which the receptor channel is closed and IP_3 is not bound. Binding of IP_3 causes the receptor to be converted to the open state, O, and we let O denote the fraction of receptors in state O. Thus the fraction of conducting receptors is O^4.

In our model we do not specifically include any binding sites for calcium. Although it appears there may be several of these on the IP_3 receptor these have not been well characterized. Because of this we instead incorporate the effects of calcium by making the rate of the S to O transition a function of calcium concentration. By doing this we are able to correlate model behavior with experimentally observed effects of calcium concentration, without having to speculate on the number or nature of calcium binding sites on the receptors. This approach has the added advantage that the reaction scheme is kept simpler.

Because of experimental evidence indicating that calcium enhances the affinity of IP_3 for its receptor we set the transition $S \to O$ to be an increasing function of calcium concentration. Thus we have assumed that calcium enhances IP_3 binding to the receptors in the model (Yoneshima *et al.*, 1997).

The transition $O \to I_1$ represents the inactivation of the receptor by calcium, while the transition $I_1 \to S$ represents the normal pathway by which the IP_3 receptor recovers from inactivation, with IP_3 dissociating from its binding site. The receptor can then rebind IP_3 and repeat the cycle. We refer to this pathway (involving S, O and I_1 only) as the "intrinsic" receptor pathway, on the basis that this reflects the direct effects of IP_3 and calcium on receptor activity. However, we also include a transition $I_1 \to I_2$, where I_2 represents a second inactivated state of the receptor in which IP_3 is no longer bound (i.e., IP_3 dissociates during this transition). This pathway is agonist-specific, and involves phosphorylation of the IP_3 receptor. Our assumption is that physiological concentrations of CCK cause rapid phosphorylation of the IP_3 receptor while physiological concentrations of ACh do not. Hence, to model the application of CCK k_4 is set to be nonzero, while the application of ACh is modeled by setting $k_4 = 0$. Phosphatase actions eventually dephosphorylate the receptor, converting it to the S state, whereupon it may rebind IP_3. A crucial element of this scheme is that while in the I_2 state the receptor is unaffected by either the IP_3 or calcium concentrations. We refer to the pathway involving all four states of the channel as the "full" pathway, with the inclusion of the generation of I_2 representing a "mediated" effect, as distinct from the intrinsic effects described above.

It is important to note that this model suffers from the flaw that it does not satisfy the principle of detailed balance. However, it is not intended as a mechanistic model of the receptor, but rather as a macroscopic description of receptor behavior. More detailed subsequent models have ensured detailed balance while retaining the same qualitative receptor behavior (Sneyd and Dufour, manuscript submitted).

By the law of mass action differential equations for the various receptor states can be determined. The equation for the fraction of receptor subunits in the open state (O) is

$$\frac{dO}{dt} = k_1 pS - k_{-1}O - k_2 O, \tag{45}$$

where p is IP_3 concentration. The variable S can be eliminated using the conservation law $S = 1 - (O + I_1 + I_2)$. Here, I_1 is the fraction of receptors in state I_1, and similarly for I_2. We assume that all four of the subunits must be in this state before the receptor will pass any calcium current. Thus the *conducting* state is O^4.

The equation for I_1 is given by

$$\frac{dI_1}{dt} = k_2 O - (k_3 + k_4) I_1, \tag{46}$$

and for I_2 by

$$\frac{dI_2}{dt} = k_4 I_1 - k_5 I_2. \tag{47}$$

As described above, the rate k_1 is a function of calcium concentration, which is denoted by c:

$$k_1 = \frac{\alpha_1 c^3}{\beta_1^3 + c^3}, \tag{48}$$

whereas k_{-1}, k_2, k_3 and k_5 are constants. The constant α_1 is the maximum rate of the S to O transition, and β_1 is the calcium concentration at which the rate of the S to O transition is half its maximum rate. The function k_1 is an increasing function of c, representing our assumption that calcium enhances the affinity of IP_3 binding.

Since we have not suggested explicit mechanisms representing the transitions governed by k_1, k_2 and k_3, we use the simplest possible functions. Thus, we choose k_2 and k_3 to be constants, and let k_1 be a sigmoidal function of c, with a Hill coefficient of 3. The rate constant k_4 is given by

$$k_4 = \frac{\alpha_4 p}{\beta_4 + p}, \tag{49}$$

and models the action of a kinase on the IP_3 receptor, and is assumed to be dependent on the type of agonist used to initiate calcium oscillations. The constant α_4 denotes the maximum rate of the reaction attained at high p, while β_4 denotes

Parameter	Value
k_{flux}	28 μM· s^{-1}
J_{influx}	0.4 (0.025) μM· s^{-1}
V_p	2.6 (0.2) μM· s^{-1}
K_p	0.54 μM
α_1	40 μM^{-1}· s^{-1}
β_1	0.8 μM
k_{-1}	0.88 s^{-1}
k_2	0.5 s^{-1}
k_3	0.5 s^{-1}
α_4	0 (0.06) s^{-1}
β_4	0.01 μM
k_5	0.02 s^{-1}

TABLE 4. Parameter values for the pancreatic acinar cell model. The values in brackets are those used to model CCK-induced oscillations.

the value of p at which the rate is half maximal. When activated, k_4 shunts the receptors through the I_2 pathway. We assume that k_4 is an increasing saturating function of the IP_3 concentration. β_4 is chosen to be relatively small, so that the rate of phosphorylation saturates quickly, and, in most physiological regimes, is approximately constant.

Finally, we incorporate the receptor model into a whole-cell model, in a fashion analogous to the other models discussed here.

The equation for the calcium concentration (c) is given by

$$\frac{dc}{dt} = J_{\text{rel}} - J_{\text{pump}} + J_{\text{influx}}, \tag{50}$$

where J_{rel} represents the IP_3-induced release of calcium from the ER and is given by

$$J_{\text{rel}} = k_{\text{flux}} O^4, \tag{51}$$

in which k_{flux} is the maximum rate of calcium release from the ER.

J_{pump} represents the activity of the calcium pumps that remove calcium from the cytosol and is given by

$$J_{\text{pump}} = \frac{V_p c^2}{K_p^2 + c^2}, \tag{52}$$

where V_p represents the maximum rate of pumping and K_p is the calcium concentration for half-maximal pumping. This expression is the same as that seen in previous models.

J_{influx} is an adjustable parameter that represents a constant rate of calcium influx into the cytosol. This term includes both influx from outside the cell and any leak of calcium from intracellular stores such as the ER and mitochondria.

The standard parameter values are given in Table 5.1.1. The values in brackets correspond to those values used to model CCK-induced oscillations, as described in more detail below.

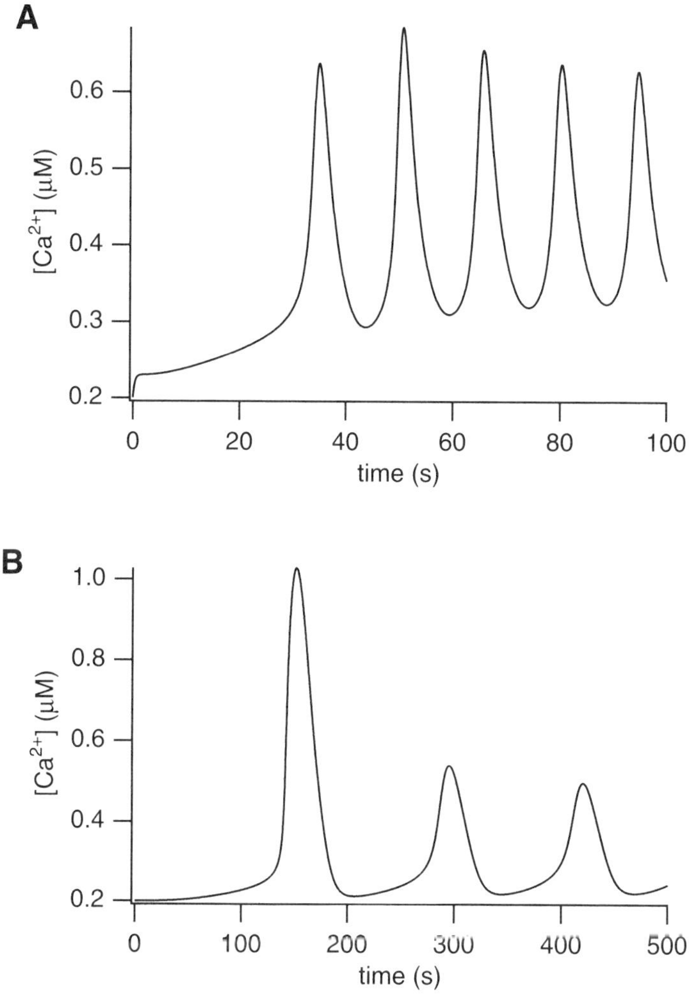

FIGURE 9. Typical oscillations of the model of calcium oscillations in pancreatic acinar cells. A: when the phosphorylation pathway is turned off (i.e., $k_4 = 0$, corresponding to the agonist ACh), and B: when the phosphorylation pathway is operative

5.1.2. *Model results, predictions, and experimental tests.* Once the model is constructed, our first job is to test whether or not such a reaction scheme can reproduce the different types of oscillations observed in pancreatic acinar cells. In Fig. 9 we show typical oscillations for two cases; A: when the phosphorylation pathway is turned off (i.e., $k_4 = 0$, corresponding to the agonist ACh), and B: when the phosphorylation pathway is operative. It is clear that in the second case the oscillations that occur have a much longer period and are spikier in shape, which is in agreement with the experimental observations.

Thus fortified by a model which predicts that such an explanation is at least possible, we then turned to experiments to confirm the model predictions. Obviously, the most important thing to check is whether or not CCK does actually cause increased levels of phosphorylation of the IP_3 receptor. The details of the experimental methods, and the detailed results, are all given in LeBeau *et al.* (1999). Suffice it to say here that we were able to show that physiological concentrations of CCK do cause increased phosphorylation of the IP_3 receptor, while that is not true of ACh.

We then used the model to investigate the relationship between the rate of calcium influx and generation of ACh-induced calcium oscillations. By reducing J_{influx} we can simulate the effect of reducing the extracellular calcium concentration. It turns out that, as J_{influx} is decreased, the range of IP_3 concentrations for which oscillations are observed shifts to the right, i.e., to higher IP_3 concentrations. Thus, a decrease in calcium influx would be expected to abolish oscillations, while an increase in IP_3 concentration would be expected to restore them. Even more interestingly, in the model a decrease in calcium influx can sometimes lead to oscillations of greatly increased period, as demonstrated in Figure 10A. The values for J_{influx} and IP_3 concentration are indicated at the top of the figure. At the normal J_{influx} value of 0.4 μMs^{-1}, oscillations are induced when the concentration of IP_3 is raised to 0.66 μM. Reducing J_{influx} to 0.35 μMs^{-1} changes the nature of the oscillations, but the original oscillatory type is recovered when the IP_3 concentration is increased to 0.75 μM.

Experiments were conducted to test these model predictions (Figure 10B). In the majority of experiments a decrease in extracellular calcium concentration led to the abolition of the oscillations, as seen in the model simulations. However, in a small number of cases it led to oscillations of greatly increased period, as shown in the figure.

5.1.3. *Conclusions.* In this example, the presence of a working model was enough to stimulate a series of experiments that were designed specifically to test the model's predictions. It is possible that these experiments might have been performed anyway at some stage, but testing for phosphorylation is a relatively long and arduous procedure. If not for the model saying it should work, I suspect that the experiments might have been long delayed. Furthermore, once the phosphorylation prediction was confirmed, we were then motivated to run further experiments to test a much broader range of model predictions. Of course, some worked out, and others didn't. One particularly important prediction that we have not yet tested is that CCK causes phosphorylation not only of the IP_3 receptor, but also the calcium ATPase pumps. There is some evidence already that this is so, but we await further confirmation from current experiments. The bottom line is that with a combination of modelling and experimental work we have learned something important about the properties of pancreatic acinar cells.

5.2. Intercellular calcium waves. My second example is a study of intercellular calcium waves that was performed over a number of years with Michael Sanderson, Andrew Charles and their colleagues (Sneyd *et al.*, 1994, 1995a).

In many cell types, a mechanical stimulus (for instance, poking a single cell with a micropipette) can initiate a wave of increased intracellular calcium that spreads from cell to cell to form an intercellular wave (Sanderson *et al.*, 1994). The epithelial cell culture forms a thin layer of cells, connected by gap junctions.

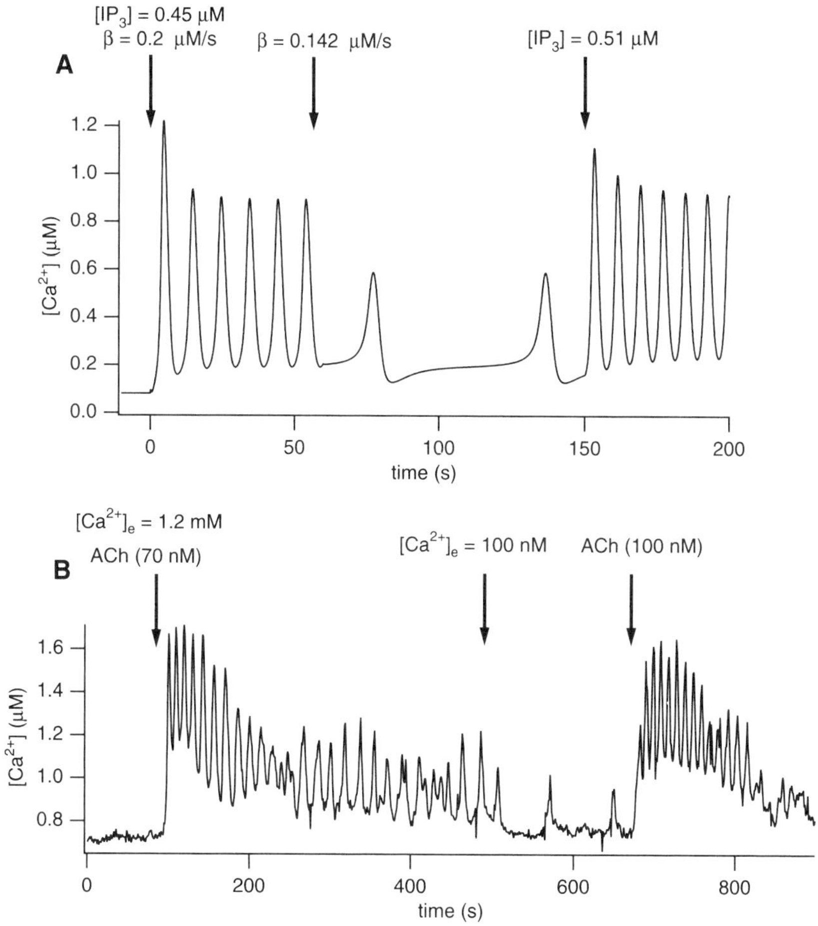

FIGURE 10. Experimental test of a theoretical prediction. A decrease in external calcium concentration can lead to oscillation of much longer period, while an increase in ACh can restore the original oscillations.

When a cell in the middle of the culture is mechanically stimulated the calcium in the stimulated cell increases quickly. After a time delay of a second or so, the neighbors of the stimulated cell also show an increase in calcium, and this increase spreads sequentially through the culture. An intracellular wave moves across each cell, is delayed at the cell boundary, and then initiates an intracellular wave in the neighboring cell. The intercellular wave moves via the sequential propagation of intracellular waves. Of particular interest here is the fact that, in the absence of extracellular calcium, the stimulated cell shows no response, but an intercellular wave still spreads to other cells in the culture. It thus appears that a rise in calcium in the stimulated cell is not necessary for wave propagation. Neither is a rise in calcium sufficient to initiate an intercellular wave. For example, epithelial cells in culture sometimes exhibit spontaneous intracellular calcium oscillations, and these

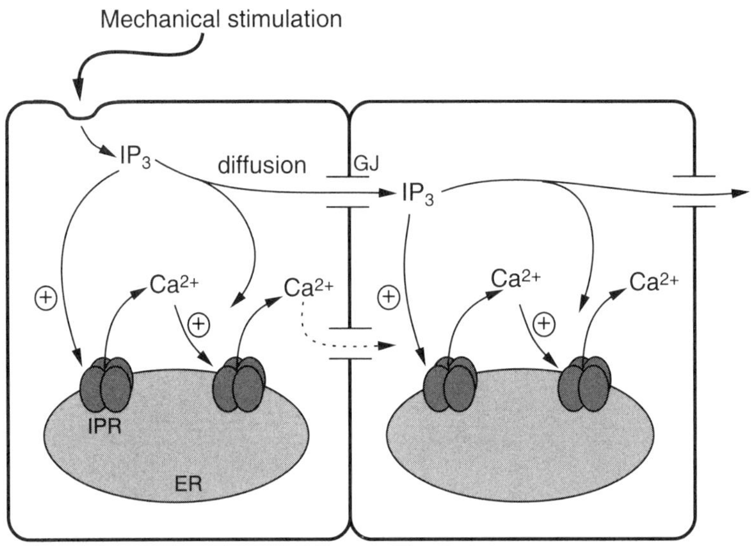

FIGURE 11. Schematic diagram of the model of intercellular calcium waves. Mechanical stimulation causes the production of IP_3, which diffuses through the cytoplasm, and from cell to cell through gap junctions (GJ). By binding to IP_3 receptors (IPR) IP_3 releases calcium from the endoplasmic reticulum (ER). This released calcium can cause the release of further calcium from the ER by activating the IPRs, as indicated by the circled + signs. However, propagation of the intercellular calcium wave is due almost entirely to the passive diffusion of IP_3 from the stimulated cell.

oscillations do not spread from cell to cell. Nevertheless, a mechanically stimulated intercellular wave does spread through cells that are spontaneously oscillating.

Little is known about the physiological importance of these intercellular calcium waves, although educated guesses can be made. Airway epithelial cells have cilia, whose function is to move mucus along the trachea. The rate at which cilia beat is closely related to the concentration of intracellular calcium in the cell; as the calcium concentration goes up, so does the ciliary beat frequency. Hence, the intercellular wave could be one way in which a group of ciliated cells coordinate a rise in beat frequency over the entire group. This would have the advantage that, if one cell is mechanically stimulated by a foreign object in the airway, it could initiate faster ciliary beating in that cell and all its neighbors, which would serve to clear the airway more efficiently. However, despite the plausibility of this hypothesis, it remains conjectural. In nonciliated cells such as glial cells, or endothelial cells, the physiological purpose of the calcium wave is even less clear.

When this behavior was first discovered, a qualitative model of the underlying mechanisms was proposed (Boitano *et al.*, 1992; Sanderson *et al.*, 1994). Sanderson and his colleagues proposed that mechanical stimulation causes the production of large amounts of IP_3 in the stimulated cell, and this IP_3 moves through the culture by passive diffusion, moving from cell to cell through gap junctions. Since IP_3 releases calcium from the endoplasmic reticulum the diffusion of IP_3 from cell to cell results in a corresponding intercellular calcium wave. This is not a wave in the mathematical sense of the word (i.e., it is not an invariant profile propagating at a constant speed), but it looks just like a wave, and it is futile to try and persuade experimentalists not to call it one. I know, I've tried. Furthermore, a number of experimental results indicated that the movement of calcium between cells does not play a major role in wave propagation. However, this qualitative model left a number of open questions. Most importantly, was this qualitative model able to account in quantitative detail for the observed properties of the waves? How fast must IP_3 diffuse in order to give waves of the correct speed? How sensitive must the IP_3 receptors be? How permeable must the gap junctions be in order to allow passage of a wave? Given appropriate numbers for the above parameters, will the wave even behave correctly at all? Will it travel in a step-wise fashion, or smoothly? Can we use a model to design other experiments that can be used to test this qualitative model?

Clearly, these quantitative questions can be answered only by the study of a mathematical model.

5.2.1. *Model equations.* In the mathematical model, the epithelial cell culture is modelled as a square grid of square cells, illustrated in Fig. 12. It is assumed that IP_3 moves by passive diffusion and is degraded with saturable kinetics. Thus, if p denotes IP_3 concentration,

$$\frac{\partial p}{\partial t} = D_p \nabla^2 p - \frac{V_p p k_p}{k_p + p}. \tag{53}$$

When $p \ll k_p$, p decays with time constant $1/V_p$. Calcium is also assumed to move by passive diffusion, but is released from the endoplasmic reticulum (ER) by IP_3 , and pumped back into the ER by calcium ATPase pumps. The equations are

$$\frac{\partial c}{\partial t} = D_c \nabla^2 c + J_{\text{flux}} - J_{\text{pump}} + J_{\text{leak}}, \tag{54}$$

$$\tau_h \frac{dh}{dt} = \frac{k_2^2}{k_2^2 + c^2} - h, \tag{55}$$

$$J_{\text{flux}} = k_f \mu(P) h \left[b + \frac{(1-b)c}{k_1 + c} \right], \tag{56}$$

$$J_{\text{pump}} = \frac{\gamma c^2}{k_\gamma^2 + c^2}, \tag{57}$$

$$J_{\text{leak}} = \beta, \tag{58}$$

$$\mu(p) = \frac{p^3}{k_\mu^3 + p^3}. \tag{59}$$

These are essentially the same as those of the heuristic model described above, equations (42)–(44). As before, J_{flux} refers to the flux of calcium through the IP_3 receptors in the ER membrane, and is a function of p, c and a slow variable h which denotes the proportion of IP_3 receptors that have not been inactivated by calcium. Similarly, J_{pump} denotes the removal of calcium from the cytoplasm by calcium ATPases in the ER membrane. J_{leak} is an unspecified leak of calcium into the cytoplasm, either from outside the cell, or from the ER. However, there are some minor differences between the Atri model and the present one. For instance, in the present model the calcium pump that removes calcium from the cytoplasm is assumed to have a Hill coefficient of 2 to better agree with experimental evidence

k_f	=	3 μMs^{-1}	b	=	0.11
k_1	=	0.7 μM	k_2	=	0.7 μM
τ_h	=	0.2 s	k_γ	=	0.27 μM
γ	=	1 μMs^{-1}	β	=	0.15 μMs^{-1}
k_μ	=	0.01 μM	V_p	=	0.08 s^{-1}
k_p	=	1 μM	F	=	2 μms^{-1}
D_c	=	20 μm^{2}s^{-1}	D_p	=	300 μm^{2}s^{-1}

TABLE 5. Parameters of the model of intercellular calcium waves.

(Lytton *et al.*, 1992), while the activation of the IP_3 receptor, described by the term $\mu(P)$ is assumed to have a Hill coefficient of 3. Values of the model parameters are given in Table 5.2.1.

Finally, the internal boundary conditions are given in terms of the flux of IP_3 from cell to cell. We assume that the flux of IP_3 across a cell boundary is linearly proportional to the concentration difference across that boundary, with proportionality constant F. Thus,

$$\text{boundary flux} = F(p_{\text{left}} - p_{\text{right}}),$$

where p_{left} and p_{right} are the concentrations on either side of the boundary. F thus has units of distance/time. Note that the concentration of IP_3 is not constant throughout any given cell, and thus the concentration difference across the cell boundary must be calculated using the IP_3 concentrations right next to the boundary.

5.2.2. *Numerical Results.* Initially, the shaded cell in Fig. 12 (the cell in the 4,4 position) was injected with IP_3 at the rate of 0.72 μMs^{-1} for 15 seconds. The IP_3 was then allowed to diffuse from cell to cell, thereby generating an intercellular calcium wave. A typical result is shown in Fig. 13, which shows a density plot of a numerical solution of the model equations in two dimensions. An intercellular wave can be seen expanding across the grid of cells and then retreating as the IP_3 degrades. The intracellular wave speed (i.e., the speed at which the intercellular wave moves across an individual cell) decreases with distance from the stimulated cell, and the arrival time and the intercellular delay increase exponentially with distance from the stimulated cell. For the values chosen for F, ranging from 1 to 8 μms^{-1}, the model agrees well with experimental data from endothelial and epithelial cells. Although it is not possible to compare the model exclusively to data from epithelial cells, the agreement with data from other cell types indicates that the model provides a reasonable quantitative description of the intercellular wave.

5.2.3. *Conclusions.* When this model was first published it created considerable controversy; a common reaction from experimentalists was "This is impossible. IP_3 cannot diffuse so far, and that mechanism is impossible". The advantage of the model was that, when faced with comments like this, we could ask which of the model parameters or assumptions was incorrect. For unless critics could pinpoint specific errors in either the assumptions or the parameters, they would be forced to conclude that our model was not so bad after all. Since then a number of experimental groups have confirmed our model in detail in other cell types (and we ourselves have designed and performed a series of experiments which have confirmed

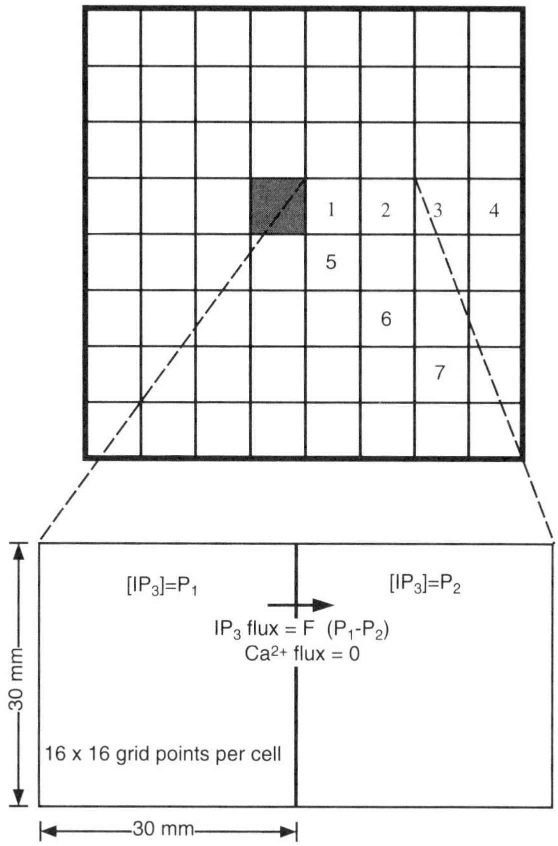

FIGURE 12. The setup for numerical simulations of the model of intercellular calcium waves. The model equations were solved on an 8×8 grid of cells, each 30 μm square. Around the perimeter of the grid, calcium and IP_3 were assumed to satisfy no-flux boundary conditions. At the internal boundaries, the flux of IP_3 between cells is proportional to the concentration difference between the cells.

the model, at least in epithelial cells). However, other experimental groups have shown that, in some cell types our model is clearly incorrect. It is now clear that there are a variety of wave propagation mechanisms, including the diffusion of extracellular messengers, the intercellular diffusion of both IP_3 and calcium, and regeneration of the signal, and these mechanisms occur with differing importance in different cell types. But our model, although now known to be wrong in detail in many cell types, and only partially correct in others, has done the job for which it was designed. It has stimulated discussion, and further experimental work.

6. A number of open questions

There are, of course, a very great number of questions that remain about the mechanisms of calcium oscillations and wave propagation. There are, equally, a

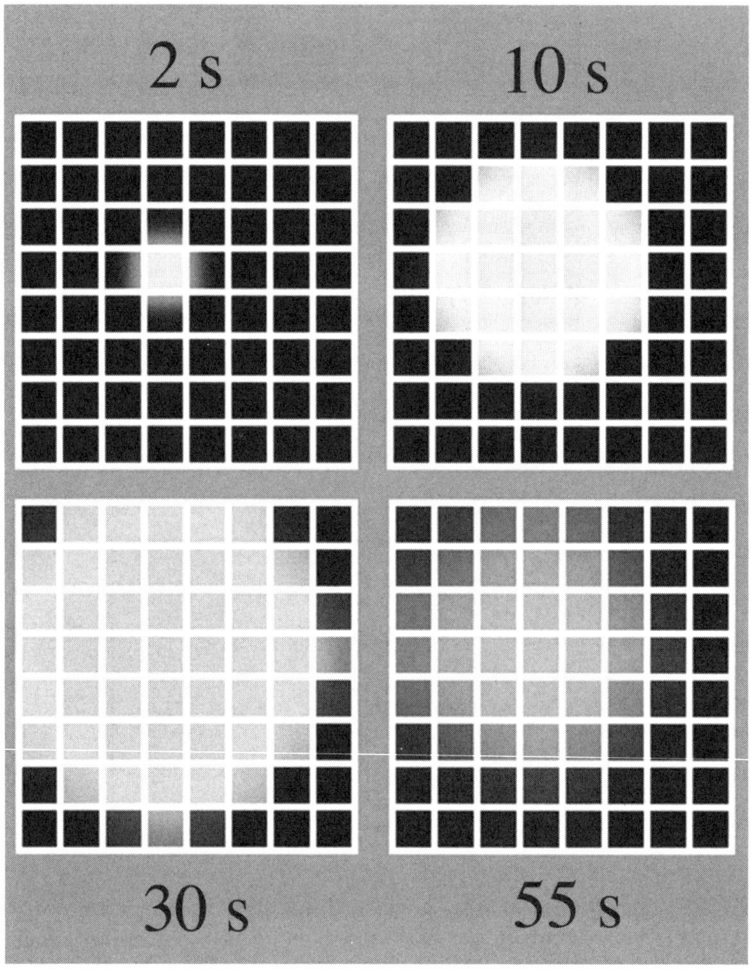

FIGURE 13. Density plot of a two-dimensional intercellular calcium wave, computed numerically from the intercellular calcium wave model.

large number of mathematical questions that have arisen from models of calcium waves. Here we have not really addressed these mathematical questions in any detail, for the simple reason that I, personally, believe them to be of less importance than questions to do with the underlying physiology. A modeler, in my opinion, should be more interested in understanding the behavior of the underlying system than in proving results about the model equations, equations which may or may not provide an acceptable description of the system under investigation. Elegant and subtle analysis of a bad model is hardly likely to produce high-quality science. I hasten to add that this point of view is not universally shared.

But although there are many questions that remain, there are some theoretical questions that stand out as particularly important, and I shall discuss these briefly.

6.1. Oscillatory intercellular waves and extracellular messengers. In some cell types (such as hepatocytes or pancreatic acinar cells), intercellular calcium waves occur in response to agonists. However, unlike the intercellular waves discussed above, these occur in a periodic manner. That is, the calcium concentration in each cell is oscillating, but these oscillations are coordinated in such a manner so as to give a periodic intercellular wave. However, we still have little understanding of the mechanisms that underlie the coordination of a large-scale intercellular wave from the interactions of each oscillating cell. Is it gap-junctional coupling by means of the diffusion of calcium through gap junctions? Or by the diffusion of IP_3 through gap junctions? As yet, we do not know for sure. Höfer (1999) has constructed a model for intercellular calcium waves in hepatocytes that shows how the diffusion of calcium through gap junctions can account for many of the observed intercellular wave properties, but Dupont *et al.* (2000) have constructed an alternate model for hepatocytes which seems to give better results when the cells are coupled by the diffusion of IP_3 from cell to cell.

Pancreatic acinar cells are ringed around a central duct, and in response to CCK, an intercellular calcium wave travels around the ring of cells. There is as yet no quantitative model for how this wave is coordinated, although Yule *et al.* (1996) have proposed a qualitative model. It appears to be governed by the diffusion of calcium through gap junctions, but this hypothesis has yet to be examined in detail.

In a variety of cell types an extracellular messenger (very probably ATP) appears to play an important role in the propagation of intercellular waves. In other words, not only is a messenger diffusing through gap junctions from cell to cell, but another messenger is diffusing in the region outside the cells, stimulating or reinforcing the calcium waves by causing the production of IP_3 in neighboring cells. The data are not always consistent, even for the same cell type, but there are a number of results that cannot be explained by appealing to passive diffusion of IP_3 through gap junctions (Charles, 1998). There have been few published quantitative studies of the interactions between extracellular and intercellular messengers, and so we do not entirely understand the types of waves to be expected from such a combination of mechanisms. For instance, is the production of ATP regenerative, and is this consistent with the observed waves? How would an intercellular wave ever stop (as is observed) if the wave propagation mechanism is made regenerative? Can one construct a partially regenerative model that is consistent in detail with the experimental evidence?

6.2. Interactions of ryanodine receptors, mitochondria and IP_3 receptors. In all the models discussed here, IP_3 receptors play the major role in calcium release. However, in many cell types (such as pancreatic acinar cells) the dynamics of calcium release are complicated by the presence of other calcium release channels such as the ryanodine receptor, and by the spatial distributions of the various calcium channels. What happens when two wave propagating mechanisms are simultaneously present in the same cell? How does this affect wave existence and speed? What happens when one of the wave mechanisms is knocked out pharmacologically? We have only a very limited understanding (from either the modelling or the experimental point of view) of these questions, and yet they are crucial for understanding calcium waves in some cell types. Mitochondria are also now known to be an important modulator of calcium oscillations and waves, but we have little understanding of their effects on model behavior (Jouaville *et* *al.*, 1995, 1998; Falcke *et al.*, 1999, 2000; Ichas *et al.*, 1997; Straub *et al.*, 2000; Colegrove *et al.*, 2000; Friel, 2000).

An additional complication is that the various types of calcium release channels have different spatial distributions within cells. For instance, in pancreatic acinar cells areas where there are high densities of IP_3 receptors are located next to areas where ryanodine receptors are situated, possibly even with a mitochondrial "barrier" in between. This spatial separation of the wave propagation mechanisms has not been modelled (at least to my knowledge) and it is not at all clear what properties are to be expected from such a situation.

6.3. Complex oscillations. In some cell types the calcium oscillations take much more complex forms, including bursting oscillations, or even chaotic behavior (Marrero *et al.*, 1994). Very little is understood about the mechanisms underlying these behaviors, although several groups have proposed models (Shen and Larter, 1995; Houart *et al.*, 1999, Kummer *et al.*, 2000; Marhl *et al.*, 2000; Haberichter *et al.*, 2001). This is likely to be one future area of great interest to both theoreticians and experimentalists.

6.4. The challenges for mathematicians. As can be clearly seen, the questions which I feel to be important are not those which are usually addressed by mathematicians, who are often more comfortable with proving analytical results about existence and uniqueness. The interpersonal and interdisciplinary skills needed for a successful collaboration are not usually included in a typical mathematician's training. To even begin to address these physiological questions, a mathematician must be willing to learn a great deal of the underlying physiology, must be prepared to go to other departments and meet the experimentalists, and must, above all, be prepared to listen to the experimental point of view while at the same time trying to express theoretical results in a manner understandable to an experimentalist.

For the mathematician who persists, the rewards are enormous.

References

[1] Atri, A., J. Amundson, D. Clapham and J. Sneyd. *A single-pool model for intracellular calcium oscillations and waves in the Xenopus laevis oocyte.* Biophys. J. **65** (1993), 1727–1739.

[2] Berne, R. M. and M. N. Levy, Eds. (1993) *Physiology.* St. Louis, Mosby Year Book.

[3] Berridge, M. J. and A. Galione. *Cytosolic calcium oscillators.* FASEB J. **2** (1988), 3074–3082.

[4] Berridge, M. J.. *Inositol trisphosphate and calcium signalling.* Nature (Lond) **361** (1993), 315–325.

[5] Berridge, M. J.. *Elementary and global aspects of calcium signalling.* J. Physiol. **499** (1997), 291–306.

[6] Bezprozvanny, I., J. Watras and B. E. Ehrlich. *Bell-shaped calcium-response curves of Ins(1,4,5)*P_3*- and calcium-gated channels from endoplasmic reticulum of cerebellum.* Nature **351** (1991), 751–754.

[7] Boitano, S., E. R. Dirksen and M. J. Sanderson. *Intercellular propagation of calcium waves mediated by inositol trisphosphate.* Science **258** (1992), 292–294.

[8] Charles, A. C., E. R. Dirksen, J. E. Merrill and M. J. Sanderson. *Mechanisms of intercellular calcium signaling in glial cells studied with dantrolene and thapsigargin.* Glia **7** (1993), 134–145.

[9] Charles, A. C.. *Glia-neuron intercellular calcium signaling.* Developmental Neuroscience **16** (1994), 196–206.

[10] Charles, A. C., S. K. Kodali and R. F. Tyndale. *Intercellular calcium waves in neurons.* Mol. Cell. Neurosci. **7** (1996), 337-53.

[11] Charles, A.. *Intercellular calcium waves in glia.* Glia **24** (1998), 39-49.

[12] Colegrove, S. L., M. A. Albrecht and D. D. Friel. *Quantitative analysis of mitochondrial Ca^{2+} uptake and release pathways in sympathetic neurons. Reconstruction of the recovery after depolarization-evoked $[Ca^{2+}]_i$ elevations.* J. Gen. Physiol. **115** (2000), 371–88.

[13] De Young, G. W. and J. Keizer. *A single pool IP_3-receptor based model for agonist stimulated Ca^{2+} oscillations.* Proc. Natl. Acad. Sci. USA **89** (1992), 9895–9899.

[14] Doedel, E., *Software for continuation and bifurcation problems in ordinary differential equations*: California Institute of Technology, 1986

[15] Duffy, M. R., N. F. Britton and J. D. Murray. *Spiral wave solutions of practical reaction-diffusion systems.* SIAM J. Appl. Math. **39** (1980), 8–13.

[16] Dupont, G. and A. Goldbeter. *One-pool model for Ca^{2+} oscillations involving Ca^{2+} and inositol 1,4,5-trisphosphate as co-agonists for Ca^{2+} release.* Cell Calcium **14** (1993), 311–322.

[17] Dupont, G. and A. Goldbeter. *Properties of intracellular Ca^{2+} waves generated by a model based on Ca^{2+}-induced Ca^{2+} release.* Biophys. J. **67** (1994), 2191–2204.

[18] Dupont, G., T. Tordjmann, C. Clair, S. Swillens, M. Claret and L. Combettes. *Mechanism of receptor-oriented intercellular calcium wave propagation in hepatocytes.* FASEB J. **14** (2000), 279–289.

[19] Endo, M., M. Tanaka and Y. Ogawa. *Calcium-induced release of calcium from the sarcoplasmic reticulum of skinned skeletal muscle fibres.* Nature **228** (1970), 34–36.

[20] Fabiato, A.. *Calcium-induced release of calcium from the cardiac sarcoplasmic reticulum.* Am. J. Physiol. **245** (1983), C1–C14.

[21] Falcke, M., J. L. Hudson, P. Camacho and J. D. Lechleiter. *Impact of mitochondrial Ca2+ cycling on pattern formation and stability..* Biophys. J. **77** (1999), 37–44.

[22] Falcke, M., M. Or-Guil and M. Bar. *Dispersion gap and localized spiral waves in a model for intracellular Ca^{2+} dynamics..* Phys. Rev. Lett. **84** (2000), 4753–6.

[23] Finch, E. A., T. J. Turner and S. M. Goldin. *Calcium as a coagonist of inositol 1,4,5-trisphosphate-induced calcium release.* Science **252** (1991), 443–446.

[24] Friel, D. D.. *Mitochondria as regulators of stimulus-evoked calcium signals in neurons..* Cell Calcium **28** (2000), 307–16.

[25] Girard, S., A. Lückhoff, J. Lechleiter, J. Sneyd and D. Clapham. *Two-dimensional model of calcium waves reproduces the patterns observed in Xenopus oocytes.* Biophys. J. **61** (1992), 509–517.

[26] Goldbeter, A., G. Dupont and M. J. Berridge. *Minimal model for signal-induced Ca^{2+} oscillations and for their frequency encoding through protein phosphorylation.* Proc. Natl. Acad. Sci. USA **87** (1990), 1461–1465.

[27] Goldbeter, A., *Biochemical oscillations and cellular rhythms: the molecular bases of periodic and chaotic behavior*: Cambridge University Press, Cambridge, 1996

[28] Guyton, A. C. and J. E. Hall, *Textbook of Medical Physiology (9th edition)*: W.B. Saunders, Philadelphia, 1996

[29] Haberichter, T., M. Marhl and R. Heinrich. *Birhythmicity, trirhythmicity and chaos in bursting calcium oscillations..* Biophys. Chem. **90** (2001), 17–30.

[30] Höfer, T.. *Model of intercellular calcium oscillations in hepatocytes: synchronization of heterogeneous cells.* Biophys. J. **77** (1999), 1244–1256.

[31] Houart, G., G. Dupont and A. Goldbeter. *Bursting, chaos and birhythmicity originating from self-modulation of the inositol 1,4,5-trisphosphate signal in a model for intracellular Ca^{2+} oscillations.* Bull. Math. Biol. **61** (1999), 507–530.

[32] Ichas, F., L. S. Jouaville and J. P. Mazat. *Mitochondria are excitable organelles capable of generating and conveying electrical and calcium signals.*. Cell **89** (1997), 1145–1153.

[33] Jaffe, L. F.. *The path of calcium in cytosolic calcium oscillations: a unifying hypothesis.* Proc. Natl. Acad. Sci. USA **88** (1991), 9883–9887.

[34] Johnston, D. and S. M.-S. Wu, *Foundations of cellular neurophysiology*: The MIT Press, Cambridge, MA, 1995

[35] Jouaville, L. S., F. Ichas, E. L. Holmuhamedov, P. Camacho and J. D. Lechleiter. *Synchronization of calcium waves by mitochondrial substrates in Xenopus laevis oocytes.* Nature **377** (1995), 438–441.

[36] Jouaville, L. S., F. Ichas and J. P. Mazat. *Modulation of cell calcium signals by mitochondria.* Mol. Cell. Biochem. **184** (1998), 371–176.

[37] Kasai, H.. *Pancreatic calcium waves and secretion.* Ciba Found. Symp. **188** (1995), 104-16.

[38] Keener, J. P.. *Waves in excitable media.* SIAM J. Appl. Math. **39** (1980), 528–548.

[39] Keener, J. P.. *A geometrical theory for spiral waves in excitable media.* SIAM J. Appl. Math. **46** (1986), 1039–1056.

[40] Keener, J. P. and J. Sneyd, *Mathematical Physiology*: Springer-Verlag, New York, 1998

[41] Keizer, J. and G. DeYoung. *Simplification of a realistic model of IP_3-induced Ca^{2+} oscillations.* J. Theor. Biol. **166** (1994), 431–442.

[42] Kopell, N. and L. N. Howard. *Plane wave solutions to reaction-diffusion equations.* Stud. Appl. Math. **52** (1973), 291–328.

[43] Kuba, K. and S. Takeshita. *Simulation of intracellular Ca^{2+} oscillation in a sympathetic neurone.* J. theor. Biol. **93** (1981), 1009–1031.

[44] Kummer, U., L. F. Olsen, C. J. Dixon, A. K. Green, E. Bornberg-Bauer and G. Baier. *Switching from simple to complex oscillations in calcium signaling.*. Biophys. J. **79** (2000), 1188–95.

[45] Lawrie, A. M., E. C. Toescu and D. V. Gallacher. *Two different spatiotemporal patterns for Ca^{2+} oscillations in pancreatic acinar cells: evidence of a role for protein kinase C in Ins(1,4,5)P_3-mediated Ca^{2+} signalling.* Cell Calcium **14** (1993), 698–710.

[46] LeBeau, A. P., D. I. Yule, G. E. Groblewski and J. Sneyd. *Agonist-dependent phosphorylation of the inositol 1,4,5-trisphosphate receptor: a possible mechanism for agonist-specific calcium oscillations in pancreatic acinar cells.* J. Gen. Physiol. **113** (1999), 851–871.

[47] Lechleiter, J., S. Girard, D. Clapham and E. Peralta. *Subcellular patterns of calcium release determined by G protein-specific residues of muscarinic receptors.* Nature **350** (1991a), 505–508.

[48] Lechleiter, J., S. Girard, E. Peralta and D. Clapham. *Spiral calcium wave propagation and annihilation in Xenopus laevis oocytes.* Science **252** (1991b), 123–126.

[49] Lechleiter, J. and D. Clapham. *Molecular mechanisms of intracellular calcium excitability in X. laevis oocytes.* Cell **69** (1992), 283–294.

[50] Li, Y.-X. and J. Rinzel. *Equations for InsP_3 receptor-mediated* $\left[\mathrm{Ca}^{2+}\right]$ *oscillations derived from a detailed kinetic model: a Hodgkin-Huxley like formalism.* J. Theor. Biol. **166** (1994), 461–473.

[51] Lytton, J., M. Westlin, S. E. Burk, G. E. Shull and D. H. MacLennan. *Functional comparisons between isoforms of the sarcoplasmic or endoplasmic reticulum family of calcium pumps.* J. Biol. Chem. **267** (1992), 14483–14489.

[52] Maginu, K.. *Geometrical characteristics associated with stability and bifurcations of periodic travelling waves in reaction-diffusion equations.* SIAM J. Appl. Math. **45** (1985), 750–774.

[53] Marhl, M., T. Haberichter, M. Brumen and R. Heinrich. *Complex calcium oscillations and the role of mitochondria and cytosolic proteins.*. Biosys. **57** (2000), 75–86.

[54] Marrero, I., A. Sanchez-Bueno, P. H. Cobbold and C. J. Dixon. *Taurolithocholate and taurolithocholate 3-sulphate exert different effects on cytosolic free Ca^{2+} concentration in rat hepatocytes.*. Biochem. J. **300** (1994), 383–6.

[55] McKenzie, A. and J. Sneyd. *On the formation and breakup of spiral waves of calcium.* Int. J. Bif. Chaos **8** (1998), 2003–2012.

[56] Murray, J. D., *Mathematical Biology*: Springer Verlag, Berlin, Heidelberg, New York, 1989

[57] Nathanson, M. H., P. J. Padfield, A. J. O'Sullivan, A. D. Burgstahler and J. D. Jamieson. *Mechanism of Ca^{2+} wave propagation in pancreatic acinar cells.* J. Biol. Chem. **267** (1992), 18118-21.

[58] Neu, J. C.. *Chemical waves and the diffusive coupling of limit cycle oscillators.* SIAM J. Appl. Math. **36** (1979), 509–515.

[59] Parker, I. and I. Ivorra. *Inhibition by Ca^{2+} of inositol trisphosphate-mediated Ca^{2+} liberation: a possible mechanism for oscillatory release of Ca^{2+}.* Proc. Natl. Acad. Sci. USA **87** (1990), 260–264.

[60] Parys, J. B., S. W. Sernett, S. DeLisle, P. M. Snyder, M. J. Welsh and K. P. Campbell. *Isolation, characterization, and localization of the inositol 1,4,5-trisphosphate receptor protein in Xenopus laevis oocytes.* J. Biol. Chem. **267** (1992), 18776–18782.

[61] Pfeiffer, F., L. Sternfeld, A. Schmid and I. Schulz. *Control of Ca^{2+} wave propagation in mouse pancreatic acinar cells.* Am. J. Physiol. **274** (1998), C663-72.

[62] Rooney, T. A. and A. P. Thomas. *Intracellular calcium waves generated by Ins(1,4,5)P_3-dependent mechanisms.* Cell Calcium **14** (1993), 674–690.

[63] Sanderson, M. J., A. C. Charles and E. R. Dirksen. *Mechanical stimulation and intercellular communication increases intracellular Ca^{2+} in epithelial cells.* Cell Reg. **1** (1990), 585–596.

[64] Sanderson, M. J., A. C. Charles, S. Boitano and E. R. Dirksen. *Mechanisms and function of intercellular calcium signaling.* Mol. Cell. Endocrin. **98** (1994), 173–187.

[65] Shen, P. and R. Larter. *Chaos in intracellular Ca^{2+} oscillations in a new model for non-excitable cells.*. Cell Calcium **17** (1995), 225–232.

[66] Sneyd, J., S. Girard and D. Clapham. *Calcium wave propagation by calcium-induced calcium release: an unusual excitable system.* Bull. Math. Biol. **55** (1993), 315–344.

[67] Sneyd, J., A. C. Charles and M. J. Sanderson. *A model for the propagation of intercellular calcium waves.* Am. J. Physiol. (Cell Physiol.) **266** (1994), C293–C302

[68] Sneyd, J., B. Wetton, A. C. Charles and M. J. Sanderson. *Intercellular calcium waves mediated by diffusion of inositol trisphosphate: a two-dimensional model.* Am. J. Physiol. (Cell Physiol.) **268** (1995a), C1537–C1545.

[69] Sneyd, J., J. Keizer and M. J. Sanderson. *Mechanisms of calcium oscillations and waves: a quantitative analysis.* FASEB J. **9** (1995b), 1463–1472.

[70] Sneyd, J., P. Dale and A. Duffy. *Traveling waves in buffered systems: applications to calcium waves.* SIAM J. Appl. Math **58** (1998), 1178–1192.

[71] Straub, S. V., D. R. Giovannucci and D. I. Yule. *Calcium wave propagation in pancreatic acinar cells: functional interaction of inositol 1,4,5-trisphosphate receptors, ryanodine receptors, and mitochondria.*. J. Gen. Physiol. **116** (2000), 547–560.

[72] Tang, Y., J. L. Stephenson and H. J. Othmer. *Simplification and analysis of models of calcium dynamics based on IP_3-sensitive calcium channel kinetics.*. Biophys. J. **70** (1996), 246–263.

[73] Thomas, A. P., G. S. J. Bird, G. Hajnóczky, L. D. Robb-Gaspers and J. W. J. Putney. *Spatial and temporal aspects of cellular calcium signaling.* FASEB J. **10** (1996), 1505–1517.

[74] Thorn, P., A. M. Lawrie, P. M. Smith, D. V. Gallacher and O. H. Petersen. *Ca^{2+} oscillations in pancreatic acinar cells: spatiotemporal relationships and functional implications.* Cell Calcium **14** (1993), 746–757.

[75] Wagner, J. and J. Keizer. *Effects of rapid buffers on Ca^{2+} diffusion and Ca^{2+} oscillations.* Biophys. J. **67** (1994), 447–456.

[76] Yoneshima, H., A. Miyawaki, T. Michikawa, T. Furuichi and K. Mikoshiba. *Ca^{2+} differentially regulates the ligand-affinity states of type 1 and type 3 inositol 1,4,5-trisphosphate receptors.* Biochem. J. **322** (1997), 591–596.

[77] Yule, D. I., A. M. Lawrie and D. V. Gallacher. *Acetylcholine and cholecystokinin induce different patterns of oscillating calcium signals in pancreatic acinar cells.* Cell Calcium **12** (1991), 145–151.

[78] Yule, D. I., E. Stuenkel and J. A. Williams. *Intercellular calcium waves in rat pancreatic acini: mechanism of transmission.* Am. J. Physiol. **271** (1996), C1285-94.

Institute of Information and Mathematical Sciences, Massey University, Albany, New Zealand

E-mail address: `j.sneyd@massey.ac.nz`

Proceedings of Symposia in Applied Mathematics
Volume **59**, 2002

Disease Gene Dynamics in a Population Isolate

Kenneth Lange and Benjamin Redelings

ABSTRACT. This paper surveys and extends mathematical methods for modeling the stochastic balance between selection and mutation in a population isolate. Geneticists are now targeting population isolates such as Iceland, Finland, and Costa Rica for mapping rare disease genes. Current population models for following the evolution of a deleterious gene involve Markov chains, branching processes, and diffusion processes. Each of these models presents interesting theoretical and numerical challenges. Here we highlight a numerical technique that gracefully bridges the gap between a discrete Markov chain and its continuous diffusion approximation. The availability of such tools should help in clarifying the issues in the debate about what populations to pursue in mapping more common genes for complex diseases.

1. Introduction

Through the early efforts of scientists such as Fisher, Haldane, Wright, and Kimura, population genetics became one of the most mathematically sophisticated branches of the biological sciences. Population genetics has served as a fertile field of application for difference and differential equations, combinatorics, Markov chains, branching processes, diffusion processes, and statistics [**2, 3, 6, 18, 19**]. In addition, population genetics has inspired new mathematics such as coalescent theory [**22**]. Despite these achievements, the attention of the majority of geneticists has focused almost entirely on molecular matters. The recent sequencing of the human genome is vivid testimony to the power and promise of molecular genetics. Nonetheless, mathematics has a curious way of reasserting itself. Closer examination of the current sequencing efforts reveals mathematics at work behind the scenes in assembling the genome. Mathematics, computer science, and statistics have also been crucial in sifting through human pedigree data in the search for disease genes.

The current paper examines some the mathematical issues that have arisen in using population isolates such as Iceland, Finland, and Costa Rica for disease gene

---

1991 *Mathematics Subject Classification.* Primary 92D25; Secondary 60J80, 60J60.

Key words and phrases. Population genetics, demography, birth-death process, Wright-Fisher Markov chain, diffusion process.

The first author was supported in part by USPHS grant GM53275 and the second author by NSF IGERT award DGE-9987641. Both authors thank Barry Merriman for helpful discussions on numerical methods for diffusion processes.

mapping. Geneticists have repeatedly turned to population isolates in mapping and cloning rare Mendelian disease genes [**17, 30, 31**]. Population isolates possess many advantages in this regard. Foremost among these is simply the fact that each isolate harbors its own trademark genetic diseases. If bank robbers go to banks because that is where the money is kept, then geneticists go to isolates because that is where the rare recessive diseases occur. Although isolates show less genetic diversity than humanity as a whole, some deleterious genes occur in them at high frequency purely by chance. Recessive disease genes also tend to persist many generations because they are subject to less stringent selection than dominant disease genes. Indeed, the majority of recessive genes occur in asymptomatic carriers and escape the selection acting on affected people. Inbreeding tends to increase the frequency of affecteds and to accelerate selection against recessive diseases.

While introduction by a subsequent mutation is possible, most recessive disease genes enter a population isolate through one of the few founders of the isolate. Genes absent in the founders never have a chance to gain a foothold. Genetic drift, the population genetics term for stochastic effects in small populations, amplifies the frequency of some genes and depresses the frequency of others. Once a recessive disease gene reaches high frequency, the low selective pressure it experiences permits it to linger for many generations. In contrast, dominant disease genes are weeded out quickly unless carriers are quite fit. Fitness, here, is a technical term referring to the ability to contribute to the next generation and not to the suffering and morbidity occasioned by the disease. Low-fitness dominant diseases are replenished by the countervailing force of mutation and tend to be more uniformly spread over the world.

Many isolates have experienced bottlenecks alternating with periods of rapid growth. Famine, war, drought, and plague are all capable of creating population bottlenecks. As an isolate rebounds from a crash, it experiences more inbreeding and the chance loss of alleles. The longer the rebound takes, the more opportunity there is for genetic drift and concomitant allele loss or amplification, particularly if population size is very low. Immigration, which is incompatible with total isolation, can restore diversity quickly if it occurs before the isolate expands dramatically in size. Mutation increases genetic diversity on a much slower time scale. Selection acts on all populations, including isolates, but its effects are more predictable in large populations.

Population geneticists have long employed branching processes to capture the extinction phenomena connected with deleterious dominant genes [**7, 8, 9, 12**]. There is also a considerable body of work modeling the stochastic balance between selection and mutation for such genes in stationary populations [**10, 11, 24, 26, 28, 34**]. A major defect of this theory is that human populations undergo systematic patterns of growth and decline incompatible with stationarity. Fortunately, the initial pessimism about revising existing branching process theory to handle non-stationary immigration was unwarranted [**15, 32**]. The key to a viable theory is the realistic assumption that new mutations occur according to a Poisson process [**4, 27**]. Constructing more complex Poisson processes by marking and then invoking Campbell's formula easily compensates for the complications of non-stationarity in the surrounding population. The current paper's restriction to the simple birth-death process [**6, 13**] allows us to illustrate the basic ideas with a minimum of branching process machinery.

For recessive diseases, the branching process paradigm of independent, noninteracting clans of carrier individuals breaks down. Selection acts only when carrier individuals of the same or different clans mate. The Wright-Fisher Markov chain serves as a substitute model for capturing the evolution of a deleterious recessive gene. Explicit formulas are hard to construct within the Wright-Fisher model, and population geneticists such as Kimura have pursued diffusion approximations to it. Even diffusion models offer limited insight for recessive diseases. There is also doubt about the applicability of the diffusion approximation in the regime of low gene frequency encountered in modeling extinction. However, the real hindrance to applying the Wright-Fisher chain and its diffusion approximation is the lack of good numerical tools. The present paper sketches a technique for smoothly bridging the gap between discrete computations based on Markov chains and continuous computations based on diffusion approximations. Our numerical scheme is motivated directly by the definition of a diffusion process and does not involve solving Kolmogorov's forward partial differential equation. We hope this scheme will prove valuable in financial modeling and other scientific applications that involve diffusion processes.

2. The Classical Selection-Mutation Model

Before launching into a discussion of stochastic models, it is helpful to review the classic deterministic model of the balance between selection and mutation for a deleterious gene [**2, 29**]. The classic model postulates discrete generations, random mating, no immigration, and a population of essentially infinite size. Within the population we consider an autosomal locus with alleles A_1 and A_2 having population frequencies p and q, respectively. For our purposes, allele A_2 is the normal allele, and allele A_1 is the deleterious allele. Mutation from A_2 to A_1 occurs at rate η; backmutation is not permitted. Denote the average fitnesses of the genotypes A_1/A_1, A_1/A_2, and A_2/A_2 by w_{A_1/A_1}, w_{A_1/A_2}, and w_{A_2/A_2}, respectively. Because only relative fitness is important, for a recessive disease, we set $w_{A_1/A_1} = f < 1$ and $w_{A_1/A_2} = w_{A_2/A_2} = 1$. For a dominant disease we set $w_{A_1/A_1} = w_{A_1/A_2} = f < 1$ and $w_{A_2/A_2} = 1$.

The frequency p_n of the disease allele A_1 for a dominant disease satisfies the recurrence

$$p_{n+1} = \frac{fp_n^2 + (1+\eta)fp_nq_n + \eta q_n^2}{fp_n^2 + f2p_nq_n + q_n^2}.$$

Given that we expect p_n to be of order η, this recurrence radically simplifies if we expand and drop all terms of order η^2 and higher. The resulting linear recurrence

$$p_{n+1} = \eta + fp_n \tag{2.1}$$

has fixed point $p_\infty = \frac{\eta}{1-f}$. Furthermore, because $p_{n+1} - p_\infty = f(p_n - p_\infty)$, the iterates p_n converge to p_∞ at linear rate f.

In contrast, the frequency p_n of the disease allele A_1 for a recessive disease satisfies the recurrence

$$p_{n+1} = \frac{fp_n^2 + (1+\eta)p_nq_n + \eta q_n^2}{fp_n^2 + 2p_nq_n + q_n^2}.$$

Now we expect p_n to be of order $\sqrt{\eta}$. Expanding the recurrence and dropping all terms of order $\eta^{3/2}$ and higher yields the quadratic recurrence

$$p_{n+1} = \eta + p_n q_n + f p_n^2 \tag{2.2}$$

with fixed point $p_\infty = \sqrt{\frac{\eta}{1-f}}$. To determine the rate of convergence, we rewrite (2.2) as

$$p_{n+1} - p_\infty = [1 - (1-f)(p_n + p_\infty)](p_n - p_\infty).$$

This makes it clear that p_n converges to p_∞ at linear rate

$$1 - (1-f)2p_\infty = 1 - 2\sqrt{\eta(1-f)}.$$

These two special cases both entail a disease prevalence on the order of η, which typically falls in the range 10^{-7} to 10^{-5}. The two cases differ markedly in their rates of convergence to equilibrium. A dominant disease quickly reaches equilibrium unless the average fitness f of affecteds is very close to 1. By comparison, a recessive diseases reach equilibrium extremely slowly.

3. Review of the Simple Birth-Death Process

The simple birth-death process is a continuous-time branching process [**6, 13**]. In a general continuous-time branching process, particles die and reproduce independently. Each particle lives an exponentially distributed length of time and, at the moment of its death, produces a random number of daughter particles. The same mean lifetime and same distribution for the number daughter particles apply to all particles. The simple birth-death process further requires that reproduction always yields either zero or two daughter particles. It helps to view the simple birth-death process as a competition between the forces of death and fission, with death having intensity μ and fission having intensity λ. The particle waits an exponentially distributed length of time with intensity $\lambda + \mu$ for either event to happen and then chooses death with probability $\mu/(\lambda + \mu)$ and fission with probability $\lambda/(\lambda + \mu)$.

The most basic questions one can ask about a simple birth-death process revolve around the random number of particles X_t alive at time t given a single particle at time 0. The probability generating function $Q(s,t) = \mathrm{E}\,(s^{X_t})$, for $s \in [0,1]$, is the most natural route to understanding the distribution of X_t. To derive an integral equation for $Q(s,t)$, we condition on the exponential lifetime T of the original particle. If we designate the distribution function of T by $F(t) = 1 - e^{-(\lambda+\mu)t}$, then it is clear that

$$Q(s,t) = \mathrm{E}\,(s^{X_t} \mid T > t)[1 - F(t)] + \int_0^t \mathrm{E}\,(s^{X_t} \mid T = u)dF(u). \tag{3.1}$$

On one hand if $T > t$, then the original particle is alive at time t and has no descendents. Hence,

$$\mathrm{E}\,(s^{X_t} \mid T > t)[1 - F(t)] = se^{-(\lambda+\mu)t}.$$

On the other hand, if the original particle dies at time $u \le t$, then it generates either zero or two daughter particles. In the latter case, the daughter particles start two separate clans that evolve independently over the remaining time span $t-u$. These considerations imply that

$$\mathrm{E}\,(s^{X_t} \mid T=u) \;=\; \frac{\mu}{\lambda+\mu}s^0 + \frac{\lambda}{\lambda+\mu}Q(s,t-u)^2.$$

Substituting these results in equation (3.1) produces

$$\begin{aligned} Q(s,t) &= se^{-(\lambda+\mu)t} + \frac{1}{\lambda+\mu}\int_0^t \left[\mu+\lambda Q(s,t-u)^2\right](\lambda+\mu)e^{-(\lambda+\mu)u}du \\ &= se^{-(\lambda+\mu)t} + \frac{\mu\left[1-e^{-(\lambda+\mu)t}\right]}{\lambda+\mu} + \lambda\int_0^t Q(s,t-u)^2e^{-(\lambda+\mu)u}du. \end{aligned}$$

This formidable integral equation for $Q(s,t)$ can be simplified by multiplying by $e^{(\lambda+\mu)t}$, regrouping terms, and changing the variable of integration from u to $w=t-u$. These steps yield the revised equation

$$\left[Q(s,t)-\frac{\mu}{\lambda+\mu}\right]e^{(\lambda+\mu)t} \;=\; s-\frac{\mu}{\lambda+\mu}+\lambda\int_0^t Q(s,w)^2e^{(\lambda+\mu)w}dw,$$

which when differentiated with respect to t gives

$$\frac{\partial}{\partial t}Q(s,t) \;=\; \lambda Q(s,t)^2-(\lambda+\mu)Q(s,t)+\mu. \tag{3.2}$$

The obvious initial condition is $Q(s,0)=s$. To simplify equation (3.2), we set $R(s,t)=1/[Q(s,t)-1]$. Straightforward calculations then give the linear differential equation

$$\frac{\partial}{\partial t}R(s,t) \;=\; (\mu-\lambda)R(s,t)-\lambda$$

with solution

$$R(s,t) \;=\; \frac{e^{(\mu-\lambda)t}}{s-1}+\frac{\lambda[e^{(\mu-\lambda)t}-1]}{\lambda-\mu}$$

satisfying the initial condition $R(s,0)=1/(s-1)$. Finally, substituting the inverse relation $Q(s,t)=1+1/R(s,t)$ produces

$$Q(s,t) \;=\; \frac{\mu(s-1)-(\lambda s-\mu)e^{(\mu-\lambda)t}}{\lambda(s-1)-(\lambda s-\mu)e^{(\mu-\lambda)t}}. \tag{3.3}$$

From the explicit solution (3.3), we reap several immediate rewards. For instance, the probability that the process goes extinct before time t is

$$\Pr(X_t=0) \;=\; Q(0,t) \;=\; \frac{\mu e^{(\mu-\lambda)t}-\mu}{\mu e^{(\mu-\lambda)t}-\lambda}. \tag{3.4}$$

The mean and variance of the number of particles at time t are

$$\begin{aligned} \mathrm{E}\,(X_t) &= \frac{\partial}{\partial s}Q(1,t) \\ &= e^{(\lambda-\mu)t} \\ \mathrm{Var}(X_t) &= \frac{\partial^2}{\partial s^2}Q(1,t)+\frac{\partial}{\partial s}Q(1,t)-\left[\frac{\partial}{\partial s}Q(1,t)\right]^2 \\ &= \frac{\lambda+\mu}{\lambda-\mu}e^{(\lambda-\mu)t}\left[e^{(\lambda-\mu)t}-1\right] \end{aligned} \tag{3.5}$$

when $\lambda\neq\mu$.

4. Poisson Processes and Campbell's Formula

Poisson processes model the formation of random points in space or time [**18, 23, 33**]. The formal definition involves a random set of points falling within some measurable space Ω equipped with a σ-finite intensity measure ω. If the set of random points is denoted by Π, and A is a measurable subset of Ω, then the counting random variable

$$N_A = \#\{\Pi \cap A\}$$

is either always finite or always countably infinite. If $\omega(A) < \infty$, then N_A is finite and follows a Poisson distribution with mean $\omega(A)$. If $\omega(A) = \infty$, then N_A is infinite. To avoid random points falling on top of one another, we require that ω possess no atoms. Finally, we assume that any finite collection $\{A_1, \ldots, A_k\}$ of disjoint measurable sets of Ω generates a collection $\{N_{A_1}, \ldots, N_{A_k}\}$ of independent counting random variables.

Campbell's theorem deals with random sums of the sort

$$S = \sum_{X \in \Pi} f(X), \tag{4.1}$$

where the real-valued function $f : \Omega \to R$ is measurable and X is a generic point of Π [**23**]. One can show that the random sum (4.1) defining S converges absolutely with probability 1 if and only if

$$\int_\Omega \min\{|f(x)|, 1\} d\omega(x) < \infty. \tag{4.2}$$

To illustrate how one evaluates the moments of S, consider the special case of a simple function $f(x) = \sum_{j=1}^m c_j 1_{A_j}$ defined by a partition $A_1, \ldots, A_m$ of Ω. Because the sets A_j are disjoint, we can write

$$S = \sum_{j=1}^m c_j N_{A_j}.$$

This representation makes it clear that

$$\begin{aligned} \mathrm{E}(S) &= \sum_{j=1}^m c_j \, \mathrm{E}(N_{A_j}) \\ &= \sum_{j=1}^m c_j \int_{A_j} d\omega(x) \qquad (4.3) \\ &= \int f(x) d\omega(x). \end{aligned}$$

If we use the independence the N_{A_j} and the fact that $\mathrm{Var}(N_{A_j}) = \mathrm{E}(N_{A_j})$, then we deduce the similar formula

$$\begin{aligned} \mathrm{Var}(S) &= \sum_{j=1}^m c_j^2 \, \mathrm{E}(N_{A_j}) \\ &= \sum_{j=1}^m c_j^2 \int_{A_j} d\omega(x) \qquad (4.4) \\ &= \int_\Omega f(x)^2 d\mu(x). \end{aligned}$$

Finally, invoking the full Poisson distributions of the N_{A_j} leads to the formulas

$$\begin{aligned} \mathrm{E}\left(e^{iuS}\right) &= \exp\left\{\int [e^{iuf(x)}-1]d\omega(x)\right\} \\ \mathrm{E}\left(u^{S}\right) &= \exp\left\{\int [u^{f(x)}-1]d\omega(x)\right\} \end{aligned} \tag{4.5}$$

for the characteristic function of S and for the probability generating function of S when $f(x)$ is nonnegative and integer valued. Appropriate limit arguments establish formulas (4.3), (4.4), and (4.5) for functions $f(x)$ more general than simple functions.

In dealing with branching processes, we will exploit a device called marking for constructing one Poisson process from another [**23**]. In marking, we suppose that there is a second measurable space Γ of possible marks for each point $x \in \Omega$. For each random point $X = x \in \Pi$, a mark $y \in \Gamma$ is independently assigned by sampling from a distribution $p(x,\cdot)$ on Γ. If the distributions $p(x,\cdot)$ are compatible in the sense that $x \to p(x,B)$ is measurable for each measurable set $B \subset \Gamma$, then we get a Poisson process on the product space $\Omega \times \Gamma$ with intensity measure

$$\nu(C) = \int\int 1_C(x,y)p(x,dy)d\omega(x).$$

This marking theorem permits straightforward calculation of certain branching process moments and probabilities through the application of Campbell's theorem.

5. Applications to a Dominant Branching Process

We are now in a position to apply the Poisson process theory just sketched to the branching process generated by mutants at an autosomal dominant locus. As the surrounding population grows, it generates an increasing number of new mutants at the locus. These new mutants, or immigrants, occur randomly over time and, to a good approximation, follow a Poisson process with intensity proportional to the population size at the current time. For the sake of simplicity, let us assume exponential population growth at rate $\alpha > 0$, starting in the infinite past and ending at the present, time $t = 0$, with a population of size n_0. The half-infinite interval $\Omega = (-\infty, 0]$ is the relevant state space for the Poisson process of new mutants.

To specify the intensity measure of the process, we assume that time is measured in units of generations, that the population has experienced constant exponential growth at rate α, and that the sex ratio is 1. To determine the macroscopic parameter α from microscopic phenomena, suppose that the average woman produces d daughters over her lifespan. If we let $w(t)$ denote the number of women at time t, then the approximation

$$w(t+\Delta t) - w(t) \approx (d-1)w(t)\Delta t, \tag{5.1}$$

implies that $w(t) = n_0 e^{(d-1)t}/2$ and therefore that $\alpha = d-1$. Counting sons as well as daughters and taking into account that each child inherits two mutable chromosomes, we expect

$$4dw(t)\eta\Delta t = 2dn_0e^{\alpha t}\eta\Delta t \tag{5.2}$$

new mutants during the time interval $(t, t+\Delta t)$. Thus, the intensity measure of the Poisson immigration process has density $2dn_0e^{\alpha t}\eta$. We mark a new mutant at time $t \in \Omega$ by the number $y \in \Gamma = \mathcal{N}_0 = \{0, 1, 2, \ldots\}$ of carriers in its clan at the present time 0. The generating function of the discrete density $p(-t, y)$ of y is given explicitly by equation (3.3) for the simple birth-death process. The reader should bear in mind that some of the theoretical formulas that follow generalize beyond the simple birth-death process and beyond exponential population growth. The details of how a clan evolves but are less important than the fact that different clans evolve independently.

The marking theorem constructs an marked Poisson process on the product space $(-\infty, 0] \times \mathcal{N}_0$ with intensity measure $\nu(t, y) = 2dn_0 e^{\alpha t}\eta p(-t, y)$. There are several functions $f(t, y) : (-\infty, 0] \times \mathcal{N}_0 \to \mathcal{R}$ that define via equation (4.1) genetically interesting sums S on this marked process. Those functions pertaining to the present depend only on y.

EXAMPLE 5.1. Current Number of Carriers

If $f(t, y) = y$, then S is the current number of carriers. Equations (3.5), (4.3), and (4.4) give the mean and variance of S as

$$\begin{aligned}
\mathrm{E}(S) &= \int_{-\infty}^{0} \sum_{y=0}^{\infty} y 2dn_0 e^{\alpha t}\eta p(-t, y)dt \\
&= 2dn_0\eta \int_{-\infty}^{0} e^{\alpha t} e^{-(\lambda-\mu)t} dt \\
&= \frac{2dn_0\eta}{\alpha - \lambda + \mu} \\
\mathrm{Var}(S) &= \int_{-\infty}^{0} \sum_{y=0}^{\infty} y^2 2dn_0 e^{\alpha t}\eta p(-t, y)dt \\
&= 2dn_0\eta \int_{-\infty}^{0} e^{\alpha t}\left[e^{-2(\lambda-\mu)t}\left(\frac{\lambda+\mu}{\lambda-\mu} + 1\right) - e^{-(\lambda-\mu)t}\frac{\lambda+\mu}{\lambda-\mu}\right] dt \\
&= 2dn_0\eta\left[\frac{2\lambda}{(\alpha - 2\lambda + 2\mu)(\lambda - \mu)} - \frac{\lambda+\mu}{(\alpha - \lambda + \mu)(\lambda - \mu)}\right].
\end{aligned}$$

These answers depend on all displayed integrals converging. Thus, $E(S)$ exists when $\alpha > \lambda - \mu$, and $\mathrm{Var}(S)$ exists when $\alpha > 2(\lambda - \mu)$. The condition $\alpha > \lambda - \mu$ simply says the rate of growth of the surrounding population exceeds that of the simple birth-death process. Nothing prevents the simple birth-death process from being supercritical.

Combining equations (3.3) and (4.5) provides the expression

$$\begin{aligned}
\mathrm{E}(u^S) &= \exp\left\{2dn_0\eta \int_{-\infty}^{0} e^{\alpha t}[Q(u, -t) - 1]dt\right\} \\
&= \exp\left\{2dn_0\eta(\mu - \lambda)(s - 1)\int_{-\infty}^{0} \frac{e^{\alpha t}}{\lambda(s-1) - (\lambda s - \mu)e^{-(\mu-\lambda)t}} dt\right\}.
\end{aligned}$$

for the probability generating function of S. We can evaluate the integral over $(-\infty, 0]$ as an infinite series by the techniques introduced in the next example. Once the probability generating function $\mathrm{E}(u^S)$ is available, we extend it to the unit circle in the complex plane and extract its coefficients by the finite Fourier transform method discussed in the references [**16, 24**]. For the sake of brevity, we omit details. ∎

EXAMPLE 5.2. Number of Extant Clans

If $f(t,y) = 1_{\{y>0\}}$, then S is the number of extant clans. The sum S counts the random number of points in the region $\{(t,y) \in (-\infty, 0] \times \mathcal{N}_0 : y > 0\}$ of the marked Poisson process and therefore is Poisson distributed. In view of formula (3.4), when $\mu < \lambda$ and $\alpha > \lambda - \mu$, the mean of S is

$$\begin{aligned} \mathrm{E}(S) &= \int_{-\infty}^{0} \sum_{y=1}^{\infty} 2dn_0 e^{\alpha t} \eta p(-t,y)dt \\ &= 2dn_0\eta \int_{-\infty}^{0} e^{\alpha t}\left[1 - \frac{\mu e^{-(\mu-\lambda)t} - \mu}{\mu e^{-(\mu-\lambda)t} - \lambda}\right] dt \\ &= \frac{2dn_0\eta(\lambda-\mu)}{\lambda} \int_{-\infty}^{0} \frac{e^{\alpha t}}{1 - \frac{\mu}{\lambda}e^{-(\mu-\lambda)t}} dt \\ &= \frac{2dn_0\eta(\lambda-\mu)}{\lambda} \sum_{k=0}^{\infty} \int_{-\infty}^{0} \left(\frac{\mu}{\lambda}\right)^k e^{(\alpha+k\lambda-k\mu)t} dt \\ &= \frac{2dn_0\eta(\lambda-\mu)}{\lambda} \sum_{k=0}^{\infty} \left(\frac{\mu}{\lambda}\right)^k \frac{1}{\alpha + k\lambda - k\mu}. \end{aligned}$$

Alternatively if $\mu > \lambda$, then a similar calculation shows that

$$\mathrm{E}(S) = \frac{2dn_0\eta(\mu-\lambda)}{\mu} \sum_{k=0}^{\infty} \left(\frac{\lambda}{\mu}\right)^k \frac{1}{\alpha + (k+1)\mu - (k+1)\lambda}.$$

Because S is Poisson, the probability that the mutant gene is currently extinct amounts to nothing more than $\Pr(S=0) = e^{-\mathrm{E}(S)}$. ■

EXAMPLE 5.3. Number of Extinct Clans

If $f(t,y) = 1_{\{y=0\}}$, then S is the number of extinct clans stretching back into the infinite past. Now S is Poisson distributed because it counts the random number of points in the region $\{(t,y) \in (-\infty, 0] \times \mathcal{N}_0\text{: y} = 0\}$. To calculate the mean of S, note that the expected number of clans, both extant and extinct, is

$$\int_{-\infty}^{0} 2dn_0 e^{\alpha t}\eta dt = \frac{2dn_0\eta}{\alpha}.$$

Thus if we subtract the mean number of extant clans calculated in the previous example from $2dn_0\eta/\alpha$, then we recover the mean number of extinct clans. ■

EXAMPLE 5.4. Distribution of the Largest Clan

If $f(t,y) = 1_{\{y>k\}}$, then S is the number of clans with more than k current members. Again S follows a Poisson distribution. Its mean

$$\mathrm{E}(S) = \int_{-\infty}^{0} 2dn_0 e^{\alpha t}\eta \sum_{y>k} p(-t,y)dt$$

reduces to the sum of Laplace transforms

$$2dn_0\eta \sum_{y>k} \hat{p}(\alpha,y) = 2dn_0\eta \sum_{y>k} \int_0^{\infty} e^{-\alpha t} p(t,y)dt \tag{5.3}$$

under exponential growth at rate α. Lange and Fan [**27**] show how to calculate these transforms explicitly. To extract the distribution of the largest clan $W = \max\{Y\}$ over all random points (T, Y) of the marked Poisson process, we use

$$\Pr(W \leq k) \quad = \quad \Pr(S = 0) \quad = \quad e^{-\mathrm{E}(S)}.$$

Finally from the distribution, one can recover the mean numerically via the identity $\mathrm{E}(W) = \sum_{k=0}^{\infty} \Pr(W > k)$. ∎

6. Specific Example for a Dominant Disease

We now illustrate the above theory using the demographic parameters of Finland [**14**]. The current Finnish population of 5,000,000 people started from an estimated 1000 founders 2000 years ago. Let us suppose for the sake of argument that one generation is approximately 25 years and that the population has experienced constant exponential growth at rate $\alpha = d - 1$, where d is the number of daughters produced by a normal woman. Assuming a sex ratio of 1, the exponential growth equation $10^3 e^{\alpha 80} = 5 \times 10^6$ predicts that $\alpha = .106465$ and $d = 1.106465$.

Now consider a carrier of a dominant disease mutation. At the end of his or her lifetime, the carrier will be replaced by $2df$ children on average, assuming a fitness f and counting both daughters and sons. Of these children, df will be carriers, and df will be normal individuals, on average. In modeling the number of carriers by a simple birth-death process, we let μ be the death rate and λ be the fission rate. These parameters are determined by the constraints

$$\begin{aligned} \lambda + \mu &= 1 \\ \frac{\mu}{\lambda + \mu} 0 + \frac{\lambda}{\lambda + \mu} 2 &= 2df. \end{aligned}$$

From the perspective of the simple birth-death process, those fission events that yield two normal children are equivalent to the death of the carrier particle, and those that yield one normal child and one carrier child can be ignored. This suggests that we use the fission and death intensities

$$\begin{aligned} \lambda^* &= \frac{1}{4}\lambda = \frac{1}{4}df \\ \mu^* &= \mu + \frac{1}{4}\lambda = 1 - \frac{3}{4}df. \end{aligned}$$

TABLE 1. Statistics of the Simple Birth-Death Process

Statistic	**Value**	**Statistic**	**Value**
carrier mean	40.00	mean extinct clans	75.38
carrier variance	77.20	largest clan mean	4.16
mean extant clans	28.55		

To compute specific carrier statistics, let us set $f = .75$ and $\eta = 10^{-6}$. Because a carrier produces on averages only $df = .83$ carrier offspring, the simple birth-death process starting with a new mutation is subcritical. Table 1 gives various statistics for the overall carrier process. Comparison with the more complete results cited in [**27**] suggests that the mean number of current carriers only partially captures the information available. An alternative model based on a geometric number of progeny per female shows the same average number of current carriers, but a larger variance for the number of carriers, a smaller average number of extant carrier clans, and a bigger mean for the largest clan. Evidently, the greater variance of the geometric progeny distribution causes more variation in clan size and a smaller clan survival rate. In compensation for a smaller survival rate, the geometric model favors the formation of a few large clans.

7. Definition of a Diffusion Process

A diffusion process X_t is a continuous-time Markov process that behaves locally like Brownian motion [**1, 19**]. Its sample paths are continuous functions confined to an interval I with left endpoint a and right endpoint b. In some applications $a = -\infty$ or $b = +\infty$ is appropriate. If a is finite, then I may be either closed or open at a, and likewise at b. The process X_t is determined by the Markovian assumption and the distribution of its increments. For small s and $X_t = x$, the increment $X_{t+s} - X_t$ is approximately normally distributed with mean and variance

$$\mathrm{E}\,(X_{t+s} - X_t \mid X_t = x) = \mu(t,x)s + o(s) \tag{7.1}$$

$$\mathrm{E}\,[(X_{t+s} - X_t)^2 \mid X_t = x] = \sigma^2(t,x)s + o(s). \tag{7.2}$$

The functions $\mu(t,x)$ and $\sigma^2(t,x) \geq 0$ are called the infinitesimal mean and variance, respectively. Here the term infinitesimal variance is used rather than infinitesimal second moment because the approximation

$$\mathrm{Var}(X_{t+s} - X_t \mid X_t = x) = \mathrm{E}\,[(X_{t+s} - X_t)^2 \mid X_t = x] + o(s)$$

follows directly from approximations (7.1) and (7.2). If the infinitesimal mean and variance do not depend on time t, then the process is time-homogeneous. The choices $\mu(t,x) = 0$ and $\sigma^2(t,x) = 1$ correspond to standard Brownian motion.

8. The Wright-Fisher Model

In the Wright-Fisher model, population size is finite. Although each current population member contributes to an infinite pool of potential gametes in proportion to his or her fitness, only a finite number of gametes is randomly sampled from the pool. Mutation from the normal allele A_2 to the deleterious allele A_1 takes place during creation of the gamete pool. The neutral model of evolution neglects mutation and treats the two alleles symmetrically.

For our purposes, the population size N_m at generation m need not be constant. The primary object of interest is the frequency X_m of allele A_1 at generation m. This frequency is the ratio of the total number Y_m of A_1 alleles to the total number of genes $2N_m$. The Wright-Fisher Markov chain specifies that Y_{m+1} is binomially distributed with $2N_{m+1}$ trials and success probability $p(X_m)$ determined by the proportion $p(X_m)$ of A_1 alleles in the pool of potential gametes for generation $m+1$. In passing to a diffusion approximation, we take one generation as the unit of time and substitute

$$\begin{aligned} \mu(m, x_m) &= \mathrm{E}\,(X_{m+1} - X_m \mid X_m = x_m) \\ &= p(x_m) - x_m \end{aligned} \tag{8.1}$$

$$\begin{aligned} \sigma^2(m, x_m) &= \mathrm{Var}(X_{m+1} - X_m \mid X_m = x_m) \\ &= \frac{p(x_m)[1 - p(x_m)]}{2N_{m+1}} \end{aligned} \tag{8.2}$$

for the infinitesimal mean $\mu(t,x)$ and variance $\sigma^2(t,x)$ of the diffusion process evaluated at time $t = m$ and position $x = x_m$ [**5, 21**].

Under neutral evolution, the gamete pool probability $p(x) = x$. This formula for $p(x)$ entails no systematic tendency for either allele to expand at the expense of the other allele. For a dominant disease, $p(x) = \eta + fx$ as suggested by equation (2.1). Similarly, for a recessive disease, $p(x) = \eta + x - (1-f)x^2$ as suggested by equation (2.2). Many population geneticists substitute $p(x) = x$ in formula (8.2) defining the infinitesimal variance $\sigma^2(t,x)$. This action is justified for neutral and recessive inheritance, but less so for dominant inheritance where the allele frequency x is typically on the order of magnitude of the mutation rate η. It is also fair to point out that in the presence of inbreeding or incomplete mixing of a population, the effective population size is less than the actual population size [**2**]. For the sake of simplicity, we will ignore this evolutionary fact.

9. Numerical Methods for Diffusion Processes

It is straightforward to simulate a diffusion process X_t. The definition tells us to extend X_t to X_{t+s} by setting the increment $X_{t+s} - X_t$ equal to a normal deviate with mean $\mu(t,x)s$ and variance $\sigma^2(t,x)s$. The time increment s should be small, and each sampled normal variate should be independent. Techniques for generating random normal deviates are covered in standard texts on computational statistics and will not be discussed here [**20, 25**]. Of more concern is how to cope with a diffusion process with finite range I. Because a normally distributed random variable has infinite range, it is possible in principle to generate an increment that takes the simulated process outside I. One remedy for this problem is to take s extremely small. It also helps if the infinitesimal variance $\sigma^2(t,x)$ tends to 0 as x approaches the boundary of I. This is the case with the neutral Wright-Fisher process.

Simulation offers a crude method of finding the distribution of X_t. Simply conduct multiple independent simulations and compute a histogram of the recorded values of X_t. Although this method is neither particularly accurate nor efficient, it has the virtue of yielding simultaneously the distributions of all of the X_t involved in the simulation process. Thus, if 1000 times are sampled per simulation, then the method yields all 1000 distributions, assuming that enough computer memory is available. Much greater accuracy can be achieved by solving Kolmogorov's forward equation. The ideal of an exact solution is seldom attained in practice, even for time-homogeneous problems. However, Kolmogorov's forward equation can be solved numerically by standard techniques for partial differential equations. Here we would like to discuss a nonstandard method for finding the distribution of X_t that directly exploits the definition of a diffusion process.

This method recursively computes the distribution of X_{t_i} at n times points labeled $0 < t_1 < \cdots < t_n = t$. In the diffusion approximation to the Wright-Fisher model, it is reasonable to let $\delta t_i = t_{i+1} - t_i$ be one generation. It is also convenient to supplement these points with the initial point $t_0 = 0$. For each t_i, we would like to compute the probability that $X_{t_i} \in [a_{ij}, a_{i,j+1}]$ for $r_i + 1$ points $a_{i0} < \cdots < a_{i,r_i}$. We will say more about these mesh points later. In the meanwhile, let p_{ij} denote the probability $\Pr(X_{t_i} \in [a_{ij}, a_{i,j+1}])$ and c_{ij} the center of probability $\mathrm{E}\,(X_{t_i} \mid X_{t_i} \in [a_{ij}, a_{i,j+1}])$. Our method carries forward approximations to both of these sequences starting from an arbitrary distribution for X_0.

In passing from time t_i to time t_{i+1}, the diffusion process redistributes a certain amount of probability from interval $[a_{ij}, a_{i,j+1}]$ to interval $[a_{i+1,k}, a_{i+1,k+1}]$. Given the definition of a diffusion process and the notation $m(i,x) = x + \mu(t_i,x)\delta t_i$ and $s^2(i,x) = \sigma^2(t_i,x)\delta t_i$, the amount redistributed is approximately

$$
\begin{aligned}
& p_{ij\to i+1,k} \\
(9.1) \qquad &= \int_{a_{ij}}^{a_{i,j+1}} \frac{1}{\sqrt{2\pi s^2(i,x)}} \int_{a_{i+1,k}}^{a_{i+1,k+1}} e^{-\frac{[y-m(i,x)]^2}{2s^2(i,x)}} dy f(t_i,x)dx. \\
&= \int_{a_{ij}}^{a_{i,j+1}} \frac{1}{\sqrt{2\pi}} \int_{[a_{i+1,k}-m(i,x)]/s(i,x)}^{[a_{i+1,k+1}-m(i,x)]/s(i,x)} e^{-\frac{z^2}{2}} dz f(t_i,x)dx.
\end{aligned}
$$

(Here and in the remainder of this section the equality sign indicates approximate equality.) In similar manner, the center of probability $c_{ij\to i+1,k}$ of the redistributed probability approximately satisfies

$$
\begin{aligned}
& c_{ij\to i+1,k}p_{ij\to i+1,k} \\
(9.2) = & \int_{a_{ij}}^{a_{i,j+1}} \frac{1}{\sqrt{2\pi s^2(i,x)}} \int_{a_{i+1,k}}^{a_{i+1,k+1}} y e^{-\frac{[y-m(i,x)]^2}{2s^2(i,x)}} dy f(t_i,x)dx \\
= & \int_{a_{ij}}^{a_{i,j+1}} \frac{1}{\sqrt{2\pi}} \int_{[a_{i+1,k}-m(i,x)]/s(i,x)}^{[a_{i+1,k+1}-m(i,x)]/s(i,x)} [m(i,x)+s(i,x)z]e^{-\frac{z^2}{2}} dz f(t_i,x)dx.
\end{aligned}
$$

Given these quantities, we calculate

$$
\begin{aligned}
p_{i+1,k} &= \sum_{j=0}^{r_i-1} p_{ij\to i+1,k} \\
(9.3) \qquad c_{i+1,k} &= \frac{1}{p_{i+1,k}} \sum_{j=0}^{r_i-1} c_{ij\to i+1,k}p_{ij\to i+1,k}
\end{aligned}
$$

on the premise that X_{t_i} is certain to belong to one of the intervals $[a_{ij}, a_{i,j+1}]$.

To carry out this updating scheme, we must approximate the integrals $p_{ij\to i+1,k}$ and $c_{ij\to i+1,k}p_{ij\to i+1,k}$. If the interval $[a_{ij}, a_{i,j+1}]$ is fairly narrow, then the linear approximations

$$
\begin{aligned}
m(i,x) &= \mu_{ij0} + \mu_{ij1}x \\
(9.4) \qquad s^2(i,x) &= \sigma_{ij}^2 \\
f(t_i,x) &= f_{ij0} + f_{ij1}x
\end{aligned}
$$

should suffice for all x in the interval. The first two of these linear approximations follow directly from the diffusion model. The constants involved in the third approximation are determined by the equations

$$\begin{aligned}
p_{ij} &= \int_{a_{i,j}}^{a_{i,j+1}} (f_{ij0} + f_{ij1}x)dx \\
&= f_{ij0}(a_{i,j+1} - a_{i,j}) + \frac{1}{2} f_{ij1}(a_{i,j+1} + a_{i,j})(a_{i,j+1} - a_{i,j}) \\
c_{ij}p_{ij} &= \int_{a_{i,j}}^{a_{i,j+1}} x(f_{ij0} + f_{ij1}x)dx \\
&= \frac{1}{2} f_{ij0}(a_{i,j+1} + a_{i,j})(a_{i,j+1} - a_{i,j}) \\
&\quad + \frac{1}{3} f_{ij1}(a_{i,j+1}^2 + a_{i,j+1}a_{i,j} + a_{i,j}^2)(a_{i,j+1} - a_{i,j})
\end{aligned}$$

with inverses

$$\begin{aligned}
f_{ij0} &= \frac{2p_{ij}(2a_{ij}^2 + 2a_{ij}a_{i,j+1} + 2a_{i,j+1}^2 - 3a_{ij}c_{ij} - 3a_{i,j+1}c_{ij})}{(a_{i,j+1} - a_{i,j})^3} \\
f_{ij1} &= \frac{6p_{ij}(2c_{ij} - a_{ij} - a_{i,j+1})}{(a_{i,j+1} - a_{i,j})^3}. \qquad (9.5)
\end{aligned}$$

It is a simple calculus exercise to check that the linear density $f_{ij0} + f_{ij1}x$ is non-negative throughout the interval $(a_{i,j}, a_{i,j+1})$ if and only its center of mass c_{ij} lies in the middle third of the interval.

Given the linear approximations (9.4), we now show that the double integrals (9.1) and (9.2) reduce to expressions involving elementary functions and the standard normal distribution function

$$\Phi(x) = \frac{1}{\sqrt{2\pi}} \int_{-\infty}^{x} e^{-y^2/2} dy.$$

The later can be rapidly evaluated by either a power series or a continued fraction expansion [**25**]. It also furnishes the key to evaluating the hierarchy of special functions

$$\Phi_k(x) = \frac{1}{\sqrt{2\pi}} \int_{-\infty}^{x} y^k e^{-y^2/2} dy$$

through the integration-by-parts recurrence

$$\Phi_k(x) = -\frac{1}{\sqrt{2\pi}} x^{k-1} e^{-x^2/2} + (k-1)\Phi_{k-2}(x) \qquad (9.6)$$

beginning with $\Phi_0(x) = \Phi(x)$. We can likewise evaluate the related integrals

$$\Psi_{jk}(x) = \int_{-\infty}^{x} y^j \Phi_k(y) dy$$

via the integration by parts reduction

$$\Psi_{jk}(x) = \frac{1}{j+1} x^{j+1} \Phi_k(x) - \frac{1}{j+1} \Phi_{j+k+1}(y). \qquad (9.7)$$

Based on the definition of $\Phi(x)$, the integral (9.1) becomes

$$p_{ij \to i+1,k} = \int_{a_{ij}}^{a_{i,j+1}} \Phi\Big(\frac{z - \mu_{ij0} - \mu_{ij1}x}{\sigma_{ij}}\Big)\Big|_{a_{i+1,k}}^{a_{i+1,k+1}} (f_{ij0} + f_{ij1}x)dx,$$

and based on the recurrence (9.6), the integral (9.2) becomes

$$\begin{aligned} & c_{ij\to i+1,k}p_{ij\to i+1,k} \\ = & \int_{a_{ij}}^{a_{i,j+1}} (\mu_{ij0}+\mu_{ij1}x)\Phi\Big(\frac{z-\mu_{ij0}-\mu_{ij1}x}{\sigma_{ij}}\Big)\Big|_{a_{i+1,k}}^{a_{i+1,k+1}} (f_{ij0}+f_{ij1}x)dx \\ & - \frac{\sigma_{ij}}{\sqrt{2\pi}}\int_{a_{ij}}^{a_{i,j+1}} e^{-(z-\mu_{ij0}-\mu_{ij1}x)^2/(2\sigma_{ij}^2)}\Big|_{a_{i+1,k}}^{a_{i+1,k+1}} (f_{ij0}+f_{ij1}x)dx. \end{aligned}$$

To evaluate the one-dimensional integrals in these expressions for $p_{ij\to i+1,k}$ and $c_{ij\to i+1,k}p_{ij\to i+1,k}$, we make appropriate linear changes of variables so that $e^{-x^2/2}$ and $\Phi(x)$ appear in the integrands and then apply formulas (9.6) and(9.7) as needed. Although the details are messy, it is clear that these maneuvers reduce everything to combinations of elementary functions and the standard normal distribution function.

To summarize, the algorithm presented approximates the probability p_{ij} and center of probability c_{ij} of each interval $[a_{ij}, a_{i,j+1}]$ of a subdivision of the range I of X_t. Equation (9.5) converts these parameters into a piecewise-linear approximation to the density of the process in preparation for propagation to the next subdivision. The actual propagation of probability from an interval of the current subdivision to another interval of the next subdivision is accomplished by computing $p_{ij\to i+1,k}$ and $c_{ij\to i+1,k}p_{ij\to i+1,k}$ based on elementary functions and the standard normal distribution function. The pieces $p_{ij\to i+1,k}$ and $c_{ij\to i+1,k}$ are then reassembled into probabilities and centers of probabilities using equations (9.3).

Choice of the mesh points $a_{i0} < \cdots < a_{i,r_i}$ at time t_i is governed by several considerations. First, the probability $\Pr(X_{t_i} \notin [a_{i0}, a_{i,r_i}])$ should be negligible. Second, $\sigma^2(t_i, x)$ should be well approximated by a constant and $\mu(t_i, x)$ by a linear function on each interval $[a_{ij}, a_{i,j+1}]$. Third, the density $f(t_i, x)$ should be well approximated by a linear function on $[a_{ij}, a_{i,j+1}]$ as well. This last requirement is the hardest to satisfy in advance, but nothing prevents one from choosing the next subdivision adaptively based on the distribution of probability within the current subdivision. Adding more mesh points will improve accuracy at the expense of efficiency. Mesh points need not be uniformly spaced. It makes sense to cluster them in regions of high probability and rapid fluctuations of $f(t, x)$. Given the smoothness expected of $f(t, x)$, rapid fluctuations are unlikely.

Many of the probabilities $p_{ij\to i+1,k}$ will be negligible. We can accelerate the algorithm by computing $p_{ij\to i+1,k}$ and $c_{ij\to i+1,k}$ only for $[a_{i+1,k}, a_{i+1,k+1}]$ close to $[a_{ij}, a_{i,j+1}]$. Because the conditional increment $X_{t_{i+1}} - X_{t_i}$ is normally distributed, it is very unlikely to extend beyond a few standard deviations σ_{ij} given X_{t_i} is in $[a_{ij}, a_{i,j+1}]$. Thus, the most sensible strategy is to visit each interval $[a_{ij}, a_{i,j+1}]$ in turn and propagate probability only to those intervals $[a_{i+1,k}, a_{i+1,k+1}]$ that lie a few standard deviations to the left or right of $[a_{ij}, a_{i,j+1}]$.

10. Numerical Methods for the Wright-Fisher Process

One of the problems with the diffusion approximation to the Wright-Fisher Markov chain is that it degrades for very low allele frequencies. Because of the interest in gene extinction, this is regrettable. However in the regime of low allele frequencies, we can always fall back on the Wright-Fisher Markov chain. As population size grows, the Markov chain updates become more and more computationally demanding. The interesting issue thus becomes how to merge the Markov chain and diffusion approaches seamlessly into a single algorithm for following the evolution of an allele. Here we present one possible algorithm and apply it to understanding disease-gene dynamics in a population isolate.

The algorithm outlined in the previous section has the virtue of being posed in terms of distribution functions rather than density functions. For low allele frequencies, discreteness is inevitable, and density functions are unrealistic. In adapting the algorithm to the regime of low allele frequencies, it is useful to let

$$a_{ij} = \frac{j - \frac{1}{2}}{2N_i}$$

for $0 \leq j \leq q$ and some positive integer q. The remaining a_{ij} are distributed over the interval $[a_{iq}, 1]$ less uniformly. This tactic separates the possibility of exactly j alleles at time t_i, $0 \leq j \leq q$, from other possibilities. For $0 \leq j \leq q$, binomial sampling dictates that

$$\begin{aligned} p_{ij \to i+1,k} &= \sum_l \binom{2N_{i+1}}{l} p^l (1-p)^{2N_{i+1}-l} \\ c_{ij \to i+1,k} &= \frac{1}{p_{ij \to i+1,k}} \sum_l \binom{2N_{i+1}}{l} \frac{l}{2N_{i+1}} p^l (1-p)^{2N_{i+1}-l} \end{aligned}$$

where $p = m(i, x)$ is the gamete pool probability at frequency $x = j/(2N_i)$ and the sums occur over all l such that $l/(2N_{i+1}) \in [a_{i+1,k}, a_{i+1,k+1})$. When $0 \leq k \leq q$, it is sensible to set $c_{ij \to i+1,k} = k/(2N_{i+1})$.

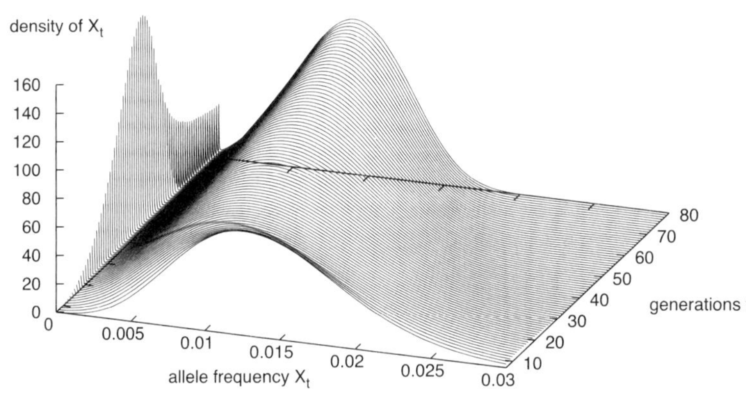

FIGURE 1. Density of the Frequency of a Recessive Gene

11. Specific Example for a Recessive Disease

To illustrate our numerical methods for a recessive disease, we again turn to Finland. Unless stated to the contrary, we assume that the population has grown exponentially from 1000 founders to $n_0 = 5,000,000$ contemporary people over a span of 80 generations. Our hypothetical recessive disease has mutation rate $\eta = 10^{-6}$, fitness $f = .5$, and initial gene frequency .015. The slow deterministic decay to the equilibrium gene frequency of $\sqrt{\eta/(1-f)} = .0014$ extends well beyond the present. Figure 1 plots the density of the frequency of the recessive gene from generation 7 to generation 80. The figure omits the first 7 generations because the densities in that time range are too concentrated for the remaining densities to scale well. The left ridge of the gene density surface represents a moderate probability mass collecting in the narrow region where the gene is either extinct or in danger of going extinct.

As a technical aside, it is interesting to compare two versions of our algorithm. Version one carries forward only probabilities and not centers of probabilities. Version two carries forward both. Version one is about twice as fast as version two, given the same mesh points at each generation. In Figure 1, version two relies on 175 intervals in the continuous region. With 2000 intervals in the continuous region, version one takes 25 times more computing cpu time and still fails to achieve the same accuracy at generation 80 as version two. Needless to say, the remaining figures in this section incorporate results from version two.

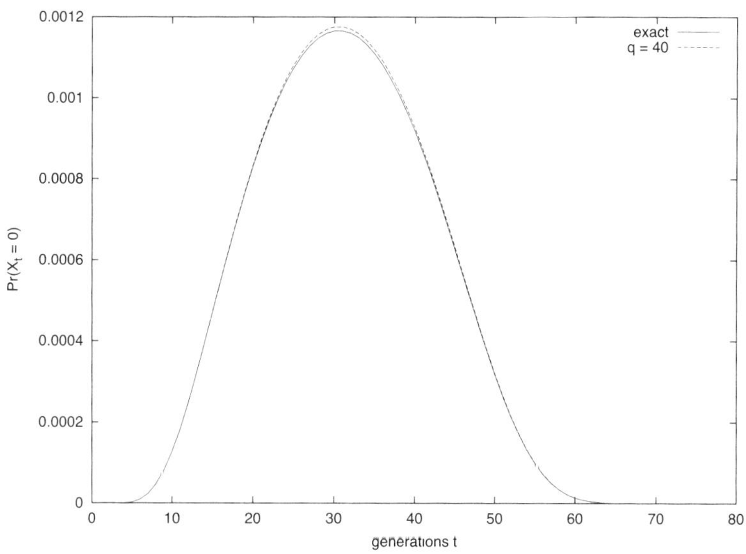

FIGURE 2. Extinction Probability of a Recessive Gene

Gene extinction is naturally of great interest. Figure 2 depicts the probability that the recessive gene is entirely absent from the population. This focuses our attention squarely on the discrete domain where we would expect the diffusion approximation to deteriorate. The solid curve of the graph shows the outcome of computing directly with the exact Wright-Fisher chain. At about generation 60, the matrix times vector multiplications implicit in the Markov chain updates start to slow the computations drastically. In this example, it took 14 minutes of computing time on a desktop PC to reach 80 generations. When we used our new algorithm with $q = 40$ intervals covering the discrete region and 500 intervals covering the continuous region, it took only 11 seconds to reach generation 80. The resulting dashed curve is quite close to the solid curve in Figure 2, and setting $q = 50$ makes it practically identical.

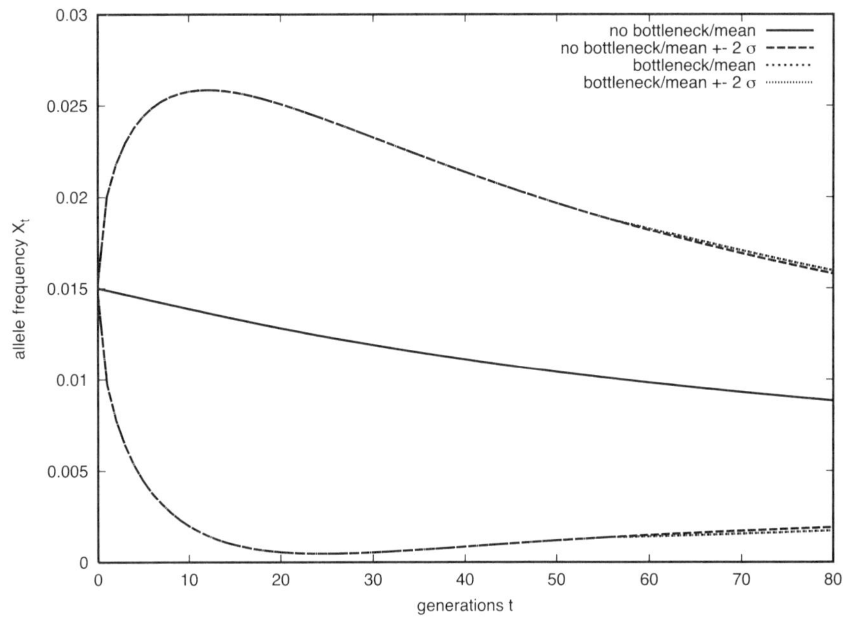

FIGURE 3. Impact of a Population Bottleneck

Figure 3 shows the impact of a population bottleneck. The conventional wisdom among mathematically naive geneticists is that bottlenecks can exert substantial influence on the frequency of disease genes. This is simply untrue unless the population dips to a very low level. The figure contrasts the means and standard deviations of the disease gene frequency computed with and without a bottleneck. To implement the bottleneck, we let the population grow exponentially from generation 0 to generation 56 to a total of 387,000 people. At generation 57 the population drops suddenly to 43,045 people and then grows exponentially thereafter to 497,444 people at generation 80, reaching $1/10$ of its current actual size. Such a precipitous drop is far in excess of what any plague or famine has caused in Finnish history, yet the differences introduced in the gene frequency distribution are barely perceptible. The mean is unaffected, as would be anticipated, and the standard deviation increases only very slightly. Less severe bottlenecks at generation 57 are unnoticeable in these plots. Of course, if a bottleneck of this fractional magnitude were introduced earlier, then the impact would be substantial.

References

[1] K. L. Chung and R. J. Williams, *Introduction to stochastic integration*, 2nd ed., Birkhäuser, Boston, MA, 1990.

[2] J. F. Crow and M. Kimura, *An introduction to population genetics theory*, Harper & Row, New York, NY, 1970.

[3] W. J. Ewens, *Mathematical population genetics*, Springer-Verlag, New York, NY, 1979.

[4] R. Fan and K. Lange, *Models for haplotype evolution in a nonstationary population*, Theor. Pop. Biol. **53** (1998), 184–198.

[5] W. Feller, *Diffusion processes in genetics*, Proceedings of the Second Berkeley Symposium on Mathematical Statistics and Probability, Univ. Calif. Press, Berkeley, CA, 1951, pp. 227–246.

[6] W. Feller, *An introduction to probability theory and its applications*, Vol. 1, 3rd ed., Wiley, New York, NY, 1968.
[7] R. A. Fisher, *The distribution of gene ratios for rare mutations*, Proc. Roy. Soc. Edinburgh, 50 (1930), 205–220.
[8] R. A. Fisher, *The genetical theory of natural selection*, Dover, New York, NY, 1958.
[9] K. Gladstien and K. Lange, *Number of people and number of generations affected by a single deleterious mutation*, Theor. Pop. Biol. **14** (1978), 313-321.
[10] K. Gladstien and K. Lange, *Equilibrium distributions for deleterious genes in large stationary populations*, Theor. Pop. Biol. **14** (1978), 322-328.
[11] J. B. S. Haldane, *A mathematical theory of natural and artificial selection, part V: selection and mutation*, Proc. Camb. Phil. Soc. **23** (1927), 838–844.
[12] J. B. S. Haldane, *The equilibrium between selection and random extinction*, Ann. Eugenics **9** (1939), 400–405.
[13] T. E. Harris, *The theory of branching processes*, Dover, New York, NY, 1989.
[14] J. Hästbacka, A. de la Chapelle, I. Kaitila, P. Sistonen, A. Weaver, and E. Lander, *Linkage disequilibrium mapping in isolated founder populations: diastrophic dysplasia in Finland*, Nature Genet. **2** (1992), 204–211.
[15] C. R. Heathcote, *A branching process allowing immigration*, J. Roy. Stat. Soc. B **27** (1965), 138–143.
[16] P. Henrici, *Fast Fourier transform methods in computational complex analysis*, SIAM Review **21** (1979), 481–527.
[17] L. B. Jorde, W. S. Watkins, J. Kere, D. Nyman, and A. W. Eriksson, *Gene mapping in isolated populations: new roles for old friends?*, Hum. Hered. **50** (2000), 57-65.
[18] S. Karlin and H. M. Taylor, *A first course in stochastic processes,* 2nd ed., Academic Press, New York, NY, 1975.
[19] S. Karlin and H, M. Taylor, *A second course in stochastic processes*, Academic Press, New York, NY, 1981.
[20] W. J. Kennedy, Jr. and J. E. Gentle, *Statistical computing*, Marcel Dekker, New York, NY, 1980.
[21] M. Kimura, *Solution of a process of random genetic drift with a continuous model*, Proc. Nat. Acad. Sci. USA **41** (1955), 144–50.
[22] J. F. C. Kingman, *The coalescent*, Stoc. Proc. Appl. **13** (1982), 235–248.
[23] J. F. C. Kingman, *Poisson Processes*, Clarendon Press, Oxford, 1993.
[24] K. Lange, *Calculation of the equilibrium distribution for a deleterious gene by the finite Fourier transform*, Biometrics **38** (1982), 79-86.
[25] K. Lange, *Numerical analysis for statisticians*. Springer-Verlag, New York, NY, 1999.
[26] K. Lange, M. Boehnke, and R. Carson, *Moment computations for subcritical branching processes*, J. Appl. Prob. **18** (1981), 52–64.
[27] K. Lange and R. Fan, *Branching process models for mutant genes in nonstationary populations*, Theor. Pop. Biol. **51** (1997), 118–133.
[28] K. Lange and K. Gladstien, *Further characterization of the long-run population distribution of a deleterious gene*, Theor. Pop. Biol. **18** (1980), 31-43.
[29] T. Nagylaki, *Introduction to theoretical population genetics*. Springer, New York, NY, 1992.
[30] L. Peltonen, *Positional cloning of disease genes: advantages of genetic isolates*, Hum. Hered. **50** (2000), 66-75.
[31] L. Peltonen, A. Palotie, and K. Lange, *The use of population isolates for mapping complex traits*, Nature Genet. Reviews **1** (2000), 182–190.
[32] M. P. Quine, *The multi-type Galton-Watson process with immigration*, J. Appl. Prob. **7** (1970), 411–422.
[33] S. M. Ross, *Stochastic processes,* 2nd ed., Wiley, New York, NY, 1996.
[34] J. G. Skellam, *The probability distribution of gene-differences in relation to selection, mutation, and random extinction*, Proc. Camb. Phil. Soc. **45** (1949), 364–367.

Departments of Biomathematics and Human Genetics, UCLA School of Medicine, Los Angeles, CA 90095-1766

E-mail address: klange@ucla.edu

Department of Biomathematics, UCLA School of Medicine, Los Angeles, CA 90095-1766

E-mail address: bredelin@ucla.edu

Proceedings of Symposia in Applied Mathematics
Volume **59**, 2002

Modeling Viral Infections

Alan S. Perelson and Patrick W. Nelson

ABSTRACT. Mathematical models have proven valuable in understanding the dynamics of viral diseases at both the population and cellular level. By comparing models to data obtained from patients initiating drug therapy, it has been possible to determine many quantitative features of HIV-1 infection, the viral infection that ultimately causes AIDS. Similar techniques are now beginning to shed light on hepatitis C and hepatitis B virus infections as well. The most dramatic finding has been that even though AIDS is a disease that occurs on a time scale of about 10 years, there are very rapid dynamical processes that occur on time scales of hours to days, as well as slower processes that occur on time scales of weeks to months that underlie disease progression. We show how dynamical modeling, using both ordinary differential equations and delay differential equations, and parameter estimation techniques have uncovered these important features of HIV-1 pathogenesis and impacted the way in which AIDS patients are treated with potent antiretroviral drugs.

1. Introduction

Mathematical models can provide insights into the dynamics of viral diseases at both the population and the cellular level. Population models began with Bernoulli in the 18th century with his study of smallpox and have been widely used since 1906 when Ross discovered the vector for malaria and attempted to model its spread. Early in the 20th century, Kermack and McKendrick studied the transmission of infectious agents and in 1927 published the first of their three classic works showing how an epidemic would spread if the density of a host is above a certain threshold [**20**]. The study of the spread of infectious diseases in a population has continued and we refer the readers to the books of Anderson [**1**], Anderson and May [**2**], Mollison [**27**], and Isham and Medley [**18**].

The application of mathematical models to the cellular aspect of infection began later and in the 1980's we began to see an explosion of models in this area. Here we will review models related to infection at the cellular level by human immunodeficiency virus-type 1 (HIV-1), the virus that causes AIDS. A new book on

1991 *Mathematics Subject Classification.* 93A30, 92B05, 92C50, 92D25.

Key words and phrases. AIDS, Mathematical Modeling, HIV, Delay Equations.

This work was performed under the auspices of the U. S. Department of Energy. Los Alamos National Laboratory strongly supports academic freedom and a researcher's right to publish; as an institution, however, the Laboratory does not endorse the viewpoint of a publication or guarantee its technical correctness.

this topic was recently written by Nowak and May [**34**]. These notes are partially derived from a extensive review of HIV models [**39**]. While not discussed in detail, the reader should be aware that these models can and have been applied to other viruses such as hepatitis C and hepatitis B virus [**31, 33, 34**].

HIV-1 infection has many puzzling quantitative features. For example, there is an average lag of nearly 10 years between infection with the virus and the onset of AIDS in adults. The reason for this time lag remains largely unknown, although during this 10 year period the number of circulating $CD4^+$ T cells slowly declines. $CD4^+$ T cells are the major target of HIV-1 infection. These cells are an important component of the immune system and when the $CD4^+$ T cell count, which is normally around 1000 mm^{-3}, reaches 200 mm^{-3}or below in an HIV-1 infected patient then that person is classified as having AIDS. Because of the central role of $CD4^+$ T cells in immune regulation, their depletion has widespread deleterious effects on the functioning of the immune system as a whole and leads to the immunodeficiency that characterizes AIDS.

The typical course of HIV-1 infection is shown in Figure 1 [**49**]. Immediately after infection the amount of virus detected in the blood, V, rises dramatically. Along with this rise in virus, flu-like symptoms tend to appear. After a few weeks to months the symptoms disappear and the virus concentration falls to a lower level. An immune response to the virus occurs and antibodies against the virus can be detected in the blood. A test to detect these antibodies is used to determine if a person has been exposed to HIV-1. If the antibodies are detected, a person is said to be HIV-1 positive.

The level the virus falls to after "primary infection" has been called the set-point. The viral concentration deviates little from this set-point level for many years; however, the concentration of $CD4^+$ T cells measured in blood slowly declines. This period, which can last for 15 years, in which the virus concentration stays relatively constant but in which the T cell count slowly falls is typically a period in which the infected person has no disease symptoms.

2. Models of primary infection

The dynamics of primary infection are much different from those observed during the asymptomatic period. After infection, there is a rapid increase in viral concentration followed in most individuals by a pronounced decline Figure 1. Around the time of peak virus concentration an immune response to the virus can be detected. This response may lead to rapid reduction in viral load but unfortunately not to viral elimination. Whether the immune response is the reason for this sharp decline in viral load is not completely understood. Phillips *et al.* [**41**] believe it is a result of a limitation in the number of cells susceptible to infection (target cell limitation), while Koup *et al.* [**21**] suggest the decline is due to an antiviral immune response. One might suspect that the decline is due to some combination of these events and possibly other factors yet to be determined. In a model by Stafford *et al.* [**45**] the viral loads observed for the first 100 days or so of primary infection could be explained by target cell limitation but in a subset of patients the degree of viral decline after the peak could not be explained unless an immune response or noncytolytic antiviral response was considered. Numerous models have been proposed to capture the dynamics of AIDS following primary infection [**17, 23, 38, 42, 45, 46**]. Most of these models focus on the dynamics of

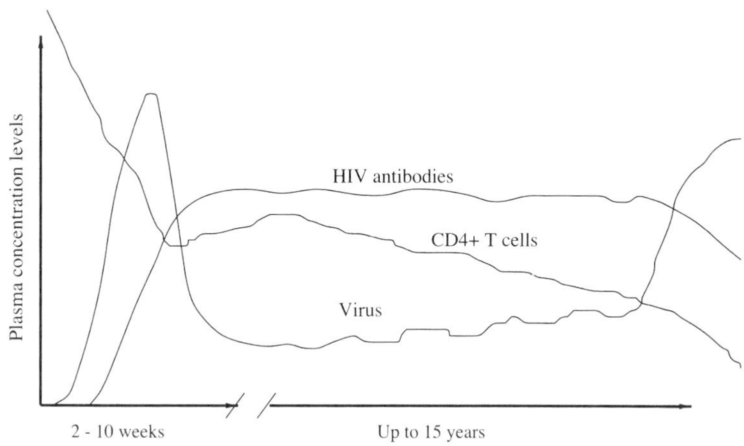

FIGURE 1. Time course of HIV-1 infection in a typical infected adult showing two distinct phases of the infection process. The first phase occurs immediately after primary infection and shows an early peak in viral load. Primary infection is followed by a long asymptomatic period during which the viral load changes little. The second phase of the infection occurs during the asymptomatic phase and in absence of drug therapy the viral load will begin to increase and eventually lead to full blown AIDS. If on the other hand the patient receives potent antiviral therapy then we may see a decline in the viral load.

target cells, infected cells and viral load and may include an effector cell response. Other models have distinguished resting and activated cells, with activated cells being targets [**9, 23, 24, 46**].

3. Models for asymptomatic infection

After the first few months of HIV-1 infection the viral load attains a set-point that remains relatively constant for years. During this period the $CD4^+$ T cell count declines slowly, but T cells remain plentiful enough to provide some immune protection. This period is called the asymptomatic period. Many investigators believed that the virus was quiescent or latent during this period. One method of determining whether the virus is active is to perturb the host-virus system during the asymptomatic period. Giving an antiretroviral drug to a patient is a means of perturbing the system away from steady state. The results of giving a protease inhibitor are shown in Figure 2; the amount of virus measured in blood plasma fell rapidly once the drug was given.

4. The simplest HIV-1 dynamic model

The data obtained from this perturbation experiment does not justify the application of a complicated model. The data appears to show that the virus concentration falls exponentially for a short period after a patient is placed on an

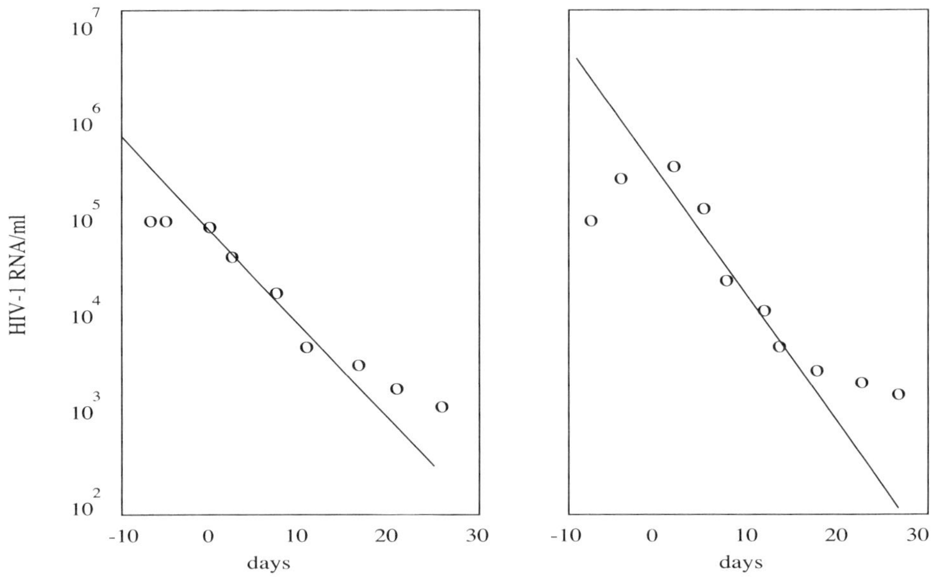

FIGURE 2. Plasma viral load before and after treatment with a protease inhibitor, showing rapid decline in viral concentrations after treatment initiation at $t = 0$. Each HIV-1 virus particle contains two RNA molecules. Data is from two out of 20 patients studied in [**16**]. All 20 patients exhibited similar rapid declines (see Table 1).

antiretroviral drug. Thus, we introduce the following model:

$$\frac{dV}{dt} = P - cV \ , \tag{1}$$

where P is an unknown function representing the rate of virus production, c is the clearance rate constant, and V is the virus concentration. If the drug completely blocks viral production, i.e., causes $P = 0$, then the model predicts V will fall exponentially, i.e., $V(t) = V_0 e^{-ct}$, where $t = 0$ is the time therapy is initiated and $V(0) = V_0$; the initial viral load. Plotting ln V vs t and using linear regression to determine the slope allowed us to estimate c, and the half-life of virus in the plasma, $t_{1/2} = ln\ 2/c$.

The notion that virus concentration attains a set-point suggests that before therapy began, the patient was in a quasi-steady state in which $dV/dt = 0$. If this were the case, as our data from these patients suggested [**16**], then by knowing c and the initial virus concentration V_0, one can compute the viral production rate before therapy, i.e., $P = cV_0$. Measuring V_0 for each patient, then multiplying this concentration by the fluid volume in which virus is expected to be found, we computed the total rate of virus production in these patients to be on the order of 12 billion viral particles per day. The rates are minimal estimates because they are based on the hypothesis that the drug completely blocks virus production and hence that the kinetics after drug therapy is initiated is a perfect exponential decline. Also, the results are based on assuming the viral dynamics behave in an instantaneous manner and hence there is no accounting for intracellular or pharmacological delays. We will first look at the dynamics when accounting for a less than perfect drug treatment and then study models that use multiple drug therapy. In the last sections we will examine models that relax all of these assumptions and include intracellular delays.

5. A model that incorporates viral production

HIV-1 infects cells that carry the $CD4^+$ cell surface protein as well as other receptors called co-receptors. Cells that are susceptible to HIV-1 infection are called target cells. The major target of HIV-1 infection is the $CD4^+$ T cell. After becoming infected such cells can produce new HIV-1 virus particles, or virions. Thus, to model HIV-1 infection we introduce a population of uninfected target cells, T, and productively infected cells, $T^\star$.

The population dynamics of $CD4^+$ T cells in humans is not well understood. Nevertheless, a reasonable model for this population of cells is

$$\frac{dT}{dt} = s + pT(1 - \frac{T}{T_{max}}) - d_T T, \tag{2}$$

where s represents the rate at which new T cells are created from sources within the body, such as the thymus. T cells can also be created by proliferation of existing T cells and we represent the proliferation by a logistic function in which p is the maximum proliferation rate and T_{max} is the T cell population density at which proliferation shuts off. Lastly, T cells, like all cells, have a natural lifespan where we define d_T as the death rate per T cell. If the population ever reaches T_{max} it should decrease; thus we impose the constraint $d_T T_{max} > s$ [**38**]. Equation (2) has two steady states but only one is biologically plausible and is given by

$$\bar{T} = \frac{T_{max}}{2p}\left[p - d_T + \sqrt{(p - d_T)^2 + \frac{4sp}{T_{max}}}\,\right], \tag{3}$$

where the over-bar denotes a steady state value. In [**38**], a discussion of biologically realistic choices for the parameters p, d_T, s, and T_{max} is given.

In the presence of HIV-1, T cells become infected. The simplest and most common method of modeling infection is to augment (2) with a "mass action" term in which the rate of infection is given by kVT, with k being the infection rate constant. Using "mass action", the rates of change of uninfected cells, T, productively infected cells, $T^\star$, and virus V, are

$$\frac{dT}{dt} = s + pT(1 - \frac{T}{T_{max}}) - d_T T - kVT, \tag{4}$$

$$\frac{dT^\star}{dt} = kVT - \delta T^\star, \tag{5}$$

$$\frac{dV}{dt} = N\delta T^\star - cV. \tag{6}$$

In the presence of HIV-1 there are two types of T cells, uninfected and productively infected. Thus, it would be reasonable to change the logistic proliferation term to $pT(1 - \frac{T+T^\star}{T_{max}})$. However, the proportion of productively infected cells is very small, on the order of 10^{-4} to 10^{-5} of T cells [**5**], and thus we ignore this correction.

Virus is produced by productively infected cells, $T^\star$. Here we have assumed that on average each productively infected cell produces N virions during its lifetime. Since the average lifetime of a productively infected cell is $1/\delta$, the average rate of virion production is $\pi = N\delta$. In some models it is useful to introduce the parameter π, while for other purposes it is easier to think about (and measure) the total number of virions produced by a cell during its lifetime, N.

In equation (6) we have ignored the loss of virus due to infection of a cell. Each time a cell is infected at least one virion must enter and thus one might add the term $-kVT$ to (6). However, if T is approximately constant then one can define a new clearance rate constant, $c' = c + kT$, that incorporates loss of virus by infection and other clearance processes.

The mechanism of virus clearance from the blood is not known. In fact, binding to cells may be an important part of the clearance mechanism. Thus, one might want to add not only a $-kVT$ term to equation (6) but similar terms for the binding to other cell populations, e.g., red blood cells, platelets, monocytes, endothelial cells, liver cells, etc. However, as long as these cell populations do not change substantially they may be considered constants and hence terms like k_1[red blood cell] + k_2 [platelet] + ... may all be lumped into the constant c.

5.1. Analysis. Before therapy is initiated, viral loads are relatively constant. Thus $dV/dt = 0$, which implies

$$N\delta T_0^\star = cV_0, \tag{7}$$

where the subscript 0 is used to denote a pre-treatment quasi-steady state value. Because V is relatively constant for weeks before therapy, this implies that $T^\star$ must also be relatively constant (assuming that the various model parameters N, c and δ are also constant). For $T^\star$ to be constant, we assume $dT^\star/dt = 0$ on this same time scale, and thus

$$kV_0T_0 = \delta T_0^\star. \tag{8}$$

Generally, T cell counts and viral loads are monitored in patients and it is reasonable to assume that the $CD4^+$ T cell concentration and the viral load are known. The vast majority of cells susceptible to HIV-1 infection are $CD4^+$ T cells [**37**] and we shall assume T_0 is equal to the $CD4^+$ T count at the start of therapy. Using equation (8) one can then determine $T_0^\star$. Thus, for patients in quasi-steady state before antiretroviral therapy begins, V_0, T_0, and $T_0^\star$ provide initial conditions for equations (4)-(6).

Equations (7) and (8) imply that for V and $T^\star$ to be in quasi-steady state,

$$NkT_0 = c. \tag{9}$$

This equation implies that patients with different quasi-steady state T cell counts must have differences in one of the parameters N, k, or c. Further, it suggests that disease progression, characterized by a lowering in the $CD4^+$ T cell count, should occur if N or k increases with time. Models incorporating these two hypotheses for disease progression have been developed in which it is assumed that within patient viral evolution drives parameter changes (cf. [**6, 35, 38, 43**]).

The T cell count changes in HIV-1 infected patients, but on a time scale of years Figure 1. If we assume that on a scale of weeks the T cell count as well as V and $T^\star$ do not change, then we can compute a full pre-treatment steady state.

This yields

$$T_0 = \frac{c}{Nk}, \quad T_0^\star = \frac{cV_0}{N\delta} \quad \text{and} \quad V_0 = \frac{sN}{c} + \frac{p - d_T}{k} - \frac{pc}{Nk^2 T_{max}}. \tag{10}$$

If we assume $T = constant = T_0$, then we can restrict our attention to

$$\begin{aligned} \frac{dT^\star}{dt} &= kT_0 V - \delta T^\star, \\ \frac{dV}{dt} &= N\delta T^\star - cV. \end{aligned} \tag{11}$$

6. Models of Drug Therapy

HIV-1 is an RNA virus. However, when it infects a cell the enzyme reverse transcriptase (RT), which it carries, makes a DNA copy of the RNA (called a provirus) and integrates it into the cell's genome. The RNA is unstable and if not transcribed into DNA permanent infection does not occur. Reverse transcriptase inhibitors block the action of RT and hence prevent infection. A second drug, a protease inhibitor, blocks a viral enzyme required for virus particles to be infectious. When protease is inhibited "non-infectious" viral particles are created. Patients who are in quasi-steady state can be given RT inhibitors, protease inhibitors or a combination of the two in order to reduce the amount of virus in their bodies. Models have been developed for all three types of therapy.

6.1. RT Inhibitors. Our model can be easily modified to account for a patient receiving an RT inhibitor. In this case there will be a reduction in k since the RT inhibitor acts to block the infectious cycle and hence there is a lesser chance of a successful interaction between an uninfected T cell and an infectious virus. For a perfect inhibitor, $k = 0$ and our basic model (4)-(6) becomes

$$\begin{aligned} \frac{dT}{dt} &= s + pT\left(1 - \frac{T}{T_{max}}\right) - d_T T, \\ \frac{dT^\star}{dt} &= -\delta T^\star, \\ \frac{dV}{dt} &= N\delta T^\star - cV. \end{aligned} \tag{12}$$

The T cell dynamic equation decouples from the viral dynamic equation. Thus, the model predicts that if viral infection has not changed any of the parameters characterizing T cell dynamics then the T cell population should eventually recover and return to its pre-infection steady state level.

Productively infected T cells are no longer generated and their number will decay exponentially, i.e., $T^\star(t) = T_0^\star e^{-\delta t}$. The amount of free virus will also decay but with more complex double exponential behavior, $V(t) = V_0 e^{-ct} + \frac{N\delta T_0^\star}{c-\delta}(e^{-\delta t} - e^{-ct})$. Assuming quasi-steady state before treatment, $T_0^\star = kV_0T_0/\delta$ and $NkT_0 = c$, yields

$$V(t) = \frac{V_0}{c - \delta}\left[ce^{-\delta t} - \delta e^{-ct}\right], \tag{13}$$

a formula presented in [**47**].

Unfortunately RT inhibitors, like other drugs, are not perfect. Thus, a more accurate model for the action of an RT inhibitor is

$$\frac{dT}{dt} = s + pT(1 - \frac{T}{T_{max}}) - d_T T - (1 - \eta_{rt})kVT, \tag{14}$$

$$\frac{dT^\star}{dt} = (1 - \eta_{rt})kVT - \delta T^\star, \tag{15}$$

$$\frac{dV}{dt} = N\delta T^\star - cV, \tag{16}$$

where η_{rt} is the "effectiveness" of the RT inhibitor. If $\eta_{rt} = 1$ the inhibition is 100% effective, whereas if $\eta_{rt} = 0$ there is no inhibition. If we assume that for a short period after therapy is initiated, $T = constant = T_0$, then equations (15) and (16) become linear and can be solved. The eigenvalues of this homogeneous linear two-dimensional system are

$$\lambda_{1,2} = -\frac{\delta + c}{2} \pm \frac{1}{2}\sqrt{(\delta + c)^2 - 4\delta[c - (1 - \eta_{rt})NkT_0]}.$$

Assuming quasi-steady state before treatment began, $NkT_0 = c$, and

$$\lambda_{1,2} = -\frac{\delta + c}{2} \pm \frac{1}{2}\sqrt{(\delta + c)^2 - 4\eta_{rt}\delta c}. \tag{17}$$

By letting $c = \epsilon\delta$, $\epsilon > 0$, it is easy to show that $(\delta + c)^2 \geq 4\delta c$, and hence since $0 < \eta_{rt} \leq 1$, the two eigenvalues are real, negative, and distinct. Thus, as $t \to \infty$, $T^\star$ and V both approach zero. Hence, if a reverse transcriptase inhibitor is given, $\eta_{rt} > 0$, and the only post-treatment steady state is the origin. Moreover, this 'uninfected' post-treatment steady state is stable. This conclusion that the virus is eradicated, however, is based on the unrealistic assumption that T cells remain constant as $t \to \infty$. In general, one expects that as the virus concentration decreases, $CD4^+$ T cells will increase in number as seen in clinical trials [**16**]. In such circumstances, η_{rt} will need to be larger than some positive critical value in order for the virus to be eliminated [**48**].

6.2. Protease Inhibitors. Protease inhibitors cause infected cells to produce noninfectious virus particles or "virions". Virions that were created prior to drug treatment remain infectious. Thus, in the presence of a protease inhibitor we consider two types of virus particles, infectious virions at concentration V_I and non-infectious virions at concentration V_{NI}. (This notation is somewhat imprecise, since even in the absence of a protease inhibitor, not every virus particle is infectious.) Let $V = V_I + V_{NI}$ be the total virus concentration.

After a 100% effective protease inhibitor is given, the equations of our basic model (4 - 6) become

$$\begin{aligned} \frac{dT}{dt} &= s + pT(1 - \frac{T}{T_{max}}) - d_T T - kV_I T, \\ \frac{dT^\star}{dt} &= kV_I T - \delta T^\star, \\ \frac{dV_I}{dt} &= (1 - \eta_p)N\delta T^\star - cV_I, \\ \frac{dV_{NI}}{dt} &= N\delta T^\star - cV_{NI} \end{aligned} \tag{18}$$

where η_p represents the effectiveness of the protease inhibitor. Before therapy is initiated, $V_{NI}(0) = 0$ and all virus belongs to the "infectious pool", i.e., $V_I(0) =$

V_0. Thus, $V_I(t) = V_0 e^{-ct}$, and as infectious virus decays the uninfected T cell population, $T(t)$, increases, ultimately returning to the steady state it had in the absence of viral infection (again assuming that infection has not caused any of the parameters characterizing the T cell population to change).

If we initially assume the protease inhibitor is completely efficacious and that $T = \text{constant} = T_0$ so we are only looking at dynamics within a week or two of the start of anti-viral therapy, one obtains a linear equation with solution

$$T^\star(t) = T^\star(0)e^{-\delta t} + \frac{kT_0V_0(e^{-ct} - e^{-\delta t})}{\delta - c}. \tag{19}$$

Assuming $T^\star$ is in quasi-steady state before initiation of therapy, $T^\star(0) = kV_0T_0/\delta$, and (19) becomes

$$T^\star(t) = kV_0T_0\left[\frac{ce^{-\delta t} - \delta e^{-ct}}{\delta(c-\delta)}\right]. \tag{20}$$

Substituting this value of $T^\star(t)$ into the differential equation for V_{NI} yields a linear time-varying equation with solution

$$V_{NI}(t) = \frac{cV_0}{c-\delta}\left[\frac{c}{c-\delta}\left(e^{-\delta t} - e^{-ct}\right) - \delta t e^{-ct}\right], \tag{21}$$

where we have used the quasi-steady state condition $NkT_0 = c$. Finally, the total concentration of virus, which is the easily measured quantity, is given by

$$V(t) = V_0e^{-ct} + \frac{cV_0}{c-\delta}\left[\frac{c}{c-\delta}\left(e^{-\delta t} - e^{-ct}\right) - \delta t e^{-ct}\right]. \tag{22}$$

Equation (22) has been used to analyze patient data [**40**]. The protease inhibitor ritonavir was given to five HIV-1 infected patients. HIV-1 RNA levels in plasma were measured after treatment at frequent intervals. Each HIV-1 virus particle contains two RNA molecules and thus the HIV-1 RNA level is a direct measure of the virus concentration, V. As shown in Figure 3, each patient responded with a similar pattern of viral decay, with an initial lag followed by an approximately exponential decline in plasma viral RNA.

Using nonlinear regression analysis, we estimated c, the viral clearance rate constant, and δ, the rate of loss of virus-producing cells, for each of the patients by fitting (22) to the plasma HIV-1 RNA measurements after a time adjustment of $t = 0$ was made to account for the pharmacological delay, i.e, the time required for drug absorption, distribution, and penetration into the target cells. The curves generated from (22), using the best-fit values of c and δ, gave a good fit to the data for all patients (Figure 3).

Clearance of free virions was the more rapid process, occurring on a time scale of hours. The values of c ranged from 2.1 to 3.8 day^{-1} with a mean of 3.1 ± 0.6 day^{-1}. The corresponding half-life, $t_{1/2}$, of free virions ($t_{1/2} = \ln 2/c$) ranged from 0.2 days to 0.3 days with a mean of 0.24 ± 0.06 days ($\sim$ 6 hours). Because data was collected every 2 hours for the first 6 hours and then every 6 hours until day 2, there were not very many data points contributing to the estimate of c, and large confidence intervals resulted [**40**]. In order to confirm that viral clearance was as rapid as predicted by this model, an additional experiment was performed in which the concentration of infectious virions in plasma, V_I, was measured in the one patient with the highest initial viral load. Theory predicted that $V_I(t) = V_0e^{-ct}$. For this patient, as shown in Figure 4, the infectivity of the patient's plasma when

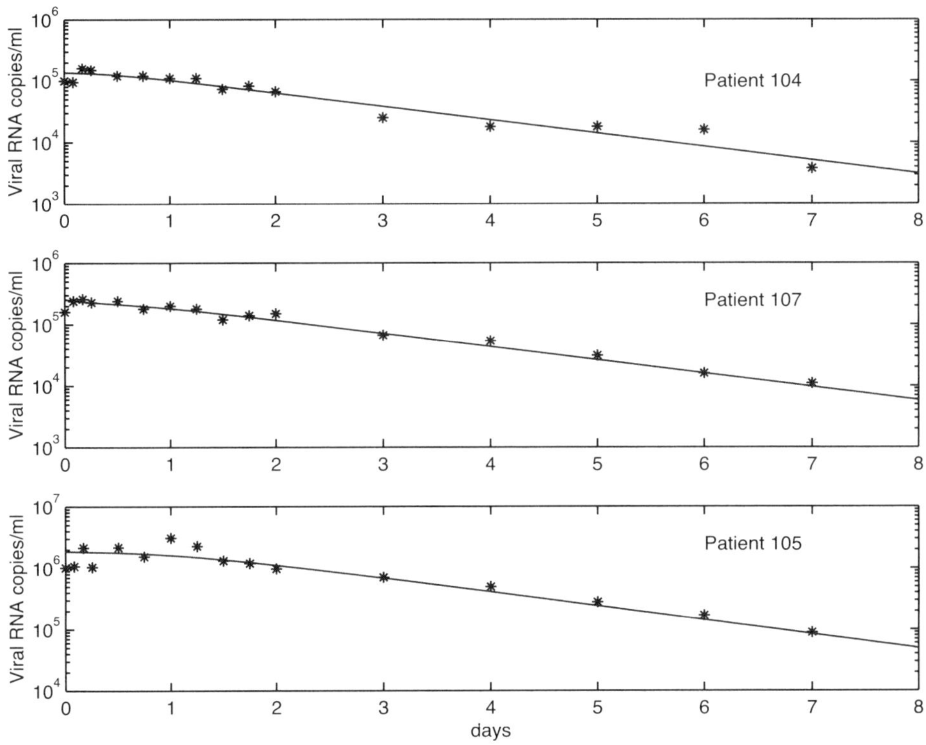

FIGURE 3. Data from three patients ($\star$) compared to the model predictions (solid lines). Parameter values corresponding to the best-fit theoretical curves are listed in Table 2.

plotted on a logarithmic scale fell linearly with time and the slope of the curve corresponded to an estimated $t_{1/2} = 0.23$ days, thus confirming the estimate of c.

The loss of virus-producing cells, as estimated from the fit of (22) to the HIV-1 RNA data, was slower than that of free virions, with values of δ ranging from 0.3 to 0.7 day^{-1}, and a mean of 0.5 ± 0.1 day^{-1}, corresponding to $t_{1/2}$ with a mean of 1.6 ± 0.6 days.

6.2.1. *Viral Production.* At steady state, the production rate of virus must equal its clearance rate, cV_0. Using the estimate of c and the pretreatment viral level, V_0, we can obtain an estimate for the rate of viral production before protease inhibitor administration. The product cV_0 gives the number of virions produced per ml of plasma per day. To compute the total number of virions produced per day we multiply cV_0 by the total fluid volume that virions are expected to be suspended in. A typical 70 kg man has a fluid volume of about 15 liters. This yields an estimate of approximately 10^{10} virions per day [**40**]. This estimate is still minimal because it only considers virus suspended in fluid and is based on the assumption that ritonavir completely blocked all infectious virus production. Additional infection and viral production is probably occurring and some virus is harbored in cells and tissues. Thus the true viral production and clearance rate is expected to be higher than our estimate.

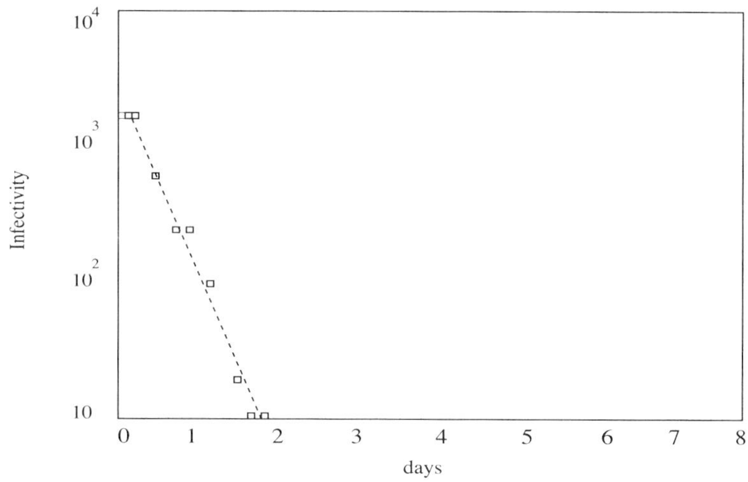

FIGURE 4. Best fit of the equation for $V_I(t)$ to plasma infectivity data obtained from patient 105. The infectivity of the plasma is proportional to the number of infectious virions it contains. See [**40**] for further experimental details.

6.2.2. *Imperfect Protease Inhibition.* Unfortunately again, protease inhibitors are not perfect drugs so we should consider (18) with $0 \le \eta_p < 1$. We can analyze this model when $T = constant = T_0$ and solve the two-dimensional linear subsystem for $T^\star$ and V_I. Assuming a pre-treatment steady state, so that $c = NkT_0$, we obtain

$$T^\star(t) = \frac{V_0 k T_0}{\lambda_2 - \lambda_1}\left[\frac{\lambda_2 + c\eta_p}{\lambda_1 + \delta}e^{\lambda_1 t} - \frac{\lambda_1 + c\eta_p}{\lambda_2 + \delta}e^{\lambda_2 t}\right], \tag{23}$$

$$V_I(t) = \frac{V_0}{\lambda_2 - \lambda_1}\left[(\lambda_2 + c\eta_p)e^{\lambda_1 t} - (\lambda_1 + c\eta_p)e^{\lambda_2 t}\right], \tag{24}$$

where

$$\lambda_{1,2} = -\frac{c+\delta}{2} \pm \frac{1}{2}\sqrt{(c+\delta)^2 - 4\eta_p \delta c}. \tag{25}$$

These eigenvalues are identical in form to those derived for the reverse transcriptase inhibitor model in Section 6.1. Thus, the eigenvalues are real, distinct and negative.

By substituting these solutions for $T^\star$ and V_I into the differential equation for V_{NI}, solving, and simplifying one finds

$$V_{NI}(t) = \frac{V_0 \eta_p}{(1-\eta_p)}\left[\frac{(\lambda_2 + c\eta_p)e^{\lambda_1 t} - (\lambda_1 + c\eta_p)e^{\lambda_2 t}}{(\lambda_2 - \lambda_1)} - e^{-ct}\right]. \tag{26}$$

Adding V_I and V_{NI}, gives the total virus

$$V(t) = \frac{V_0}{(1-\eta_p)}\left[\frac{(\lambda_2 + c\eta_p)e^{\lambda_1 t} - (\lambda_1 + c\eta_p)e^{\lambda_2 t}}{(\lambda_2 - \lambda_1)} - \eta_p e^{-ct}\right]. \tag{27}$$

If the drug effectiveness η_p is close to 1, then the $1-\eta_p$ term in the denominator can cause numerical difficulties. This singularity can be avoided by using the characteristic equation and the steady state condition $NkT_0 = c$ to replace $1-\eta_p$ with $(\lambda_1 + c)(\lambda_1 + \delta)/\delta c$.

While this solution is more complex than that given in (22), it reduces to the simpler form when $\eta_p = 1$. This can be seen by using the substitution for $1-\eta_p$ given above coupled with l'Hôpital's rule, and noting that when $\eta_p = 1$ that $\lambda_1 = -\delta$ and $\lambda_2 = -c$.

6.2.3. *Patients not in quasi-steady state.* It is not necessary to assume that patients are in quasi-steady state prior to treatment initiation, hence $c \neq NkT_0$. This assumption gives a more general characteristic equation with eigenvalues,

$$\lambda_{1,2} = -\frac{\delta + c}{2} \pm \frac{1}{2}\sqrt{(\delta - c)^2 + 4(1-\eta_p)NkT_0\delta}. \tag{28}$$

Using these eigenvalues, one can show the decay of $V(t)$ is monotonic if $c \geq NkT_0$. When $c < NkT_0$, the virus is increasing when therapy is initiated, and will continue to increase for a while under therapy, as illustrated in Figure 5.

These eigenvalues differ from (25) and provide a means for studying the model's predictions for cases when patients are not at a steady state when treatment is initiated. A bifurcation occurs at

$$c = NkT_0(1-\eta_p),$$

or equivalently at

$$\eta_p = 1 - \frac{c}{NkT_0}.$$

If $\eta_p > 1 - \dfrac{c}{NkT_0}$, then both eigenvalues are negative and the virus is predicted to be cleared from the plasma. If $\eta_p < 1 - \dfrac{c}{NkT_0}$ then therapy will not be effective and the virus will continue to grow . Notice that if $\eta_p = 1$, then, as we have already shown, the stability analysis predicts the virus will always be eliminated since the eigenvalues become $\lambda_1 = -c, \lambda_2 = -\delta, \lambda_3 = -c$. However, these predictions depend on the assumption that $T(t) = T_0$, which tends to fail since after weeks of therapy T cells increase in the vast majority of patients.

6.2.4. *Viral elimination.* Consider a patient in pre-therapy steady state. If we apply protease inhibitor therapy and allow T cells to change with time, one can predict the level of drug therapy needed for viral eradication based on the patient's initial (pre-therapy) level of CD4$^+$ T cells. One of the steady states of (18) corresponds to the virus being eliminated. This steady state is defined by $(\bar{T}, \bar{T^\star}, \bar{V_I}, \bar{V_{NI}}) = (T_{ss1}, 0, 0, 0)$, where

$$T_{ss1} = \frac{T_{max}}{2p}\left[p - d_T + \sqrt{(p-d_T)^2 + \frac{4sp}{T_{max}}}\right]. \tag{29}$$

If this steady state is stable then the model predicts that the virus will be eliminated. Analyzing (18) about this steady state gives the characteristic equation,

$$\left[p - d_T - \frac{2pT_{ss1}}{T_{max}} - \lambda\right]\left[(\lambda+\delta)(\lambda+c) - (1-\eta_p)N\delta kT_{ss1}\right] = 0. \tag{30}$$

$CD4^+$ cells	Drug efficacy
$1000/\text{mm}^3$	Healthy Patient
$500/\text{mm}^3$	$\eta_p > .5$
$200/\text{mm}^3$	$\eta_p > .8$

TABLE 1. Predicted drug efficacy needed for viral eradication for patients with varying initial levels of $CD4^+$ T cells. We have assumed a healthy patient has about $1000/mm^3$ T cells and that the patient had a pre-treatment steady state.

Studying the first bracketed term of (30), we find the conditions for the first eigenvalue to be negative is given by

$$T_{ss1} > \frac{(p - d_T)T_{max}}{2p} \tag{31}$$

which reduces to

$$\sqrt{(p - d_T)^2 + \frac{4sp}{T_{max}}} > 0. \tag{32}$$

Hence, this eigenvalue is always negative and so to determine stability we must look at the second bracketed term of (30), which is the same as (28) with T_{ss1} replacing T_0. Hence, if the drug efficacy satisfies

$$\eta_p > 1 - \frac{c}{NkT_{ss1}} \tag{33}$$

then all eigenvalues will be negative. Under the assumption of a pre-treatment steady state, $c = NkT_0$, this condition reduces to

$$\eta_p > 1 - \frac{T_0}{T_{ss1}}. \tag{34}$$

In Table 1 we show the critical drug efficacies required to eliminate the virus as a function of T_0 and the patients pre-therapy level of $CD4^+$ T cells, under the assumption $T_{ss1} = 1000\ mm^{-3}$.

6.3. The effects of T cell recovery on viral dynamics. In the previous analyses we have assumed that $T = constant = T_0$. However, after antiretroviral therapy is initiated some recovery of T cells is observed. Data obtained in [**16**] suggests that over a period of weeks the recovery of T cells can be described by either a linear or exponential function of time, with no statistically significant difference between the two functions. Thus assume

$$T(t) = T_0 + \alpha t. \tag{35}$$

so that

$$\frac{dT^\star}{dt} = kV_I(\alpha t + T_0) - \delta T^\star, \tag{36}$$

$$\frac{dV_I}{dt} = (1 - \eta_p)N\delta T^\star - cV_I, \tag{37}$$

$$\frac{dV_{NI}}{dt} = \eta_p N\delta T^\star - cV_{NI}. \tag{38}$$

Rewriting equations (36) and (37) as a second order equation in $V_I(t)$ gives

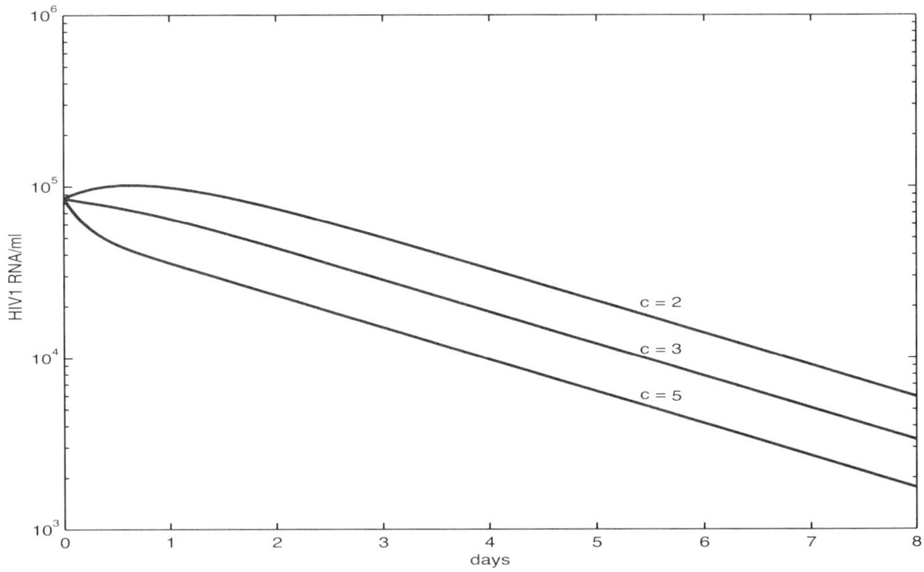

FIGURE 5. Prediction of viral decay under 100% effective protease inhibition ($\eta_p = 1$) when the assumption of a pre-treatment steady state is not made. Here $NkT_0 = 2.56$ day^{-1}, and the viral load V initially increases if the clearance rate c is less than this value.

$$\tfrac{d^2V_I}{dt^2} + (c+\delta)\tfrac{dV_I}{dt} + (\delta c - (1-\eta_p)Nk\delta(\alpha t + T_0))V_I = 0.$$

The substitution $V_I(t) = e^{bt}Y(t)$ yields

$$Y'' + 2bY' + b^2Y + (c+\delta)(bY + Y') + (\delta c - (1-\eta_p)N\delta k(\alpha t + T_0))Y = 0 \quad ,$$

which is of the form

$$Y'' + \zeta(t)\,Y = 0 \ ,$$

where

$$\zeta(t) = -\frac{(c+\delta)^2}{4} + \delta c - (1-\eta_p)\,Nk\,(\alpha t + T_0)\,\delta.$$

We can rewrite $\zeta(t)$ as $\zeta(t) = a + bt$ with a and b constants, and define a transformation of the independent variable as, $s = a + bt$, to get

$$Y_{ss} - b^2 sY = 0,$$

which is an Airy equation, whose solutions have been extensively studied. Hence using the initial conditions and numerical methods we have an explicit solution to compare with experimental data. Figure 6 shows the solution and its fit with experimental data from three patients. Interestingly, explicitly taking the change in CD4$^+$ T cells that occur over the course of one week of therapy into account has little effect on the overall change in HIV-1 RNA.

6.4. Viral Generation Time. Virologists are interested in determining how rapidly virus replicates. One means of doing this is to estimate the generation time

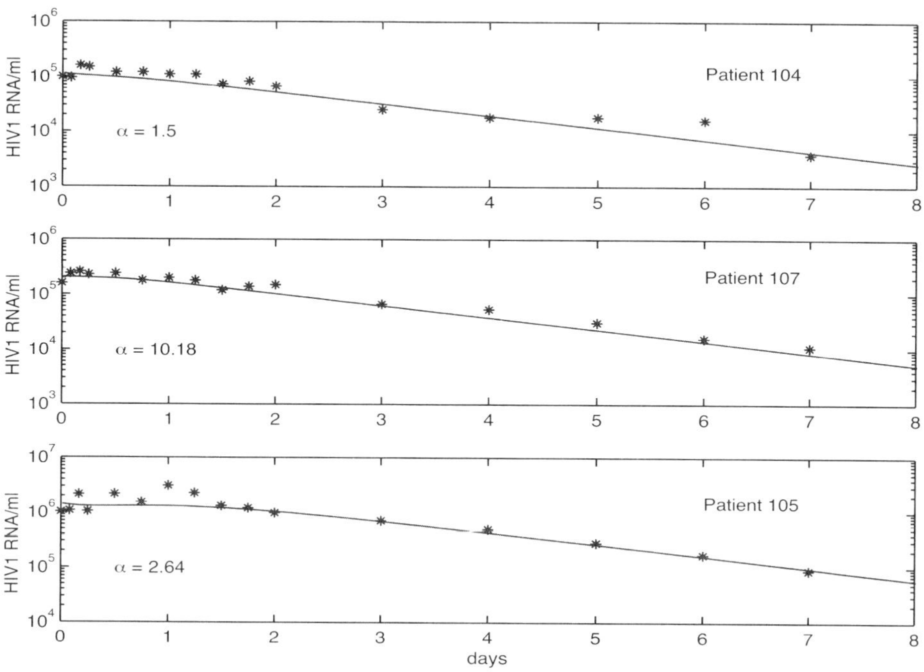

FIGURE 6. The effect of a linear increase in T on viral dynamics. Graph showing model predictions plotted with patient data. The values for α were determined by linear regression analysis of the change in CD4$^+$ T cell levels for each patient after drug treatment. The plotted curves are indistinguishable from the theoretical curves shown in Figure 3 generated assuming $T = constant = T_0$.

of virus. Here we shall define the generation time of HIV-1 as the time it takes a population of virions to infect cells and reproduce. Because a single virion may be cleared before it infects a cell, we cannot simply average the generation times of individual particles, for this average will diverge. The procedure we follow is to examine a steady-state population of V_0 virions in a patient with $T = constant = T_0$ target cells. We then call the average time for this population to be replaced by a new population of V_0 virions the generation time, T_g. Thus, for example if the original virions were all colored red, and newly produced virions were all colored blue, we compute the average time until the appearance of the new generation of V_0 blue particles. While virions are not colored, the use of a 100% effective protease inhibitor causes newly produced virions to be non-infectious, i.e., blue. Because we wish to calculate the time until V_0 non-infectious particles are created, we assume such particles are not cleared and that initially none exist. Thus, $dV_{NI}/dt = N\delta T^\star$, with $V_{NI}(0) = 0$. Further, we shall assume that at $t = 0$ there are no productively infected cells so that only new infections are tracked. Thus, $T^\star(0) = 0$, and from equation (19)

$$T^\star(t) = \frac{kT_0V_0}{c-\delta}(e^{-\delta t} - e^{-ct}) \tag{39}$$

At any given time, t, the mean number of "blue" virions produced from the initial V_0 "red" virions is $V_{NI}(t)$. If we let $P(t)$ be the (cumulative) probability that a virion is produced by time t, then $P(t) = V_{NI}(t)/V_0$. The probability density of a virus being produced at time t is $p(t) = dP/dt$, and thus the average time of virion production is given by

$$T_g = \int_0^\infty tp(t)dt = \frac{1}{V_0}\int_0^\infty t\frac{dV_{NI}}{dt}dt = \frac{1}{V_0}\int_0^\infty tN\delta T^\star dt. \tag{40}$$

Substituting the value of $T^\star(t)$ given above, using the quasi-steady state condition $NkT_0 = c$, and integrating, we find

$$T_g = \frac{1}{\delta} + \frac{1}{c}. \tag{41}$$

What we have called the viral generation time has a very simple interpretation. In the life cycle of HIV-1, virions are either free or within cells. The generation time is thus the sum of the average times spent in these two compartments, $1/c$ being the average lifetime of a free virion and $1/\delta$ being the average lifespan of an infected cell. There is also another interpretation. Because the system is at quasi-steady state, $c = NkT_0$, the clearance rate and rate of new cell infection are coupled. Thus, the viral generation time can also be viewed as the time for an infected cell to produce N new virions, i.e., its lifespan $1/\delta$, plus the time for this cohort of N virions to infect any of the T_0 uninfected target cells, i.e., $1/(NkT_0)$. For the five patients studied in [**40**], the average generation time was 2.6 days. In more recent studies employing combination therapy [**36**] or the RT inhibitor nevirapine [**13**], higher estimates of δ were obtained, which lead to a generation time of 1.8 days. This implies that in a steady-state patient HIV-1 goes through $\sim$ 140 - 200 replication cycles per year.

7. Drug Resistance

The fact that $> 10^{10}$ virions are produced each day in the average mid-stage HIV-1 infected patient has significant implications for the generation of drug resistance. When HIV-1 replicates, its RNA genome is reverse transcribed into DNA. This copying process is error-prone, and in vivo the error rate has been estimated as 3×10^{-5} per base per replication cycle [**22**]. The HIV-1 genome has approximately 10^4 bases, and thus the average number of changes per genome is 0.3 per replication cycle. According to the binomial distribution or its Poisson approximation, we then expect after one replication 74% of infected cells to carry nonmutated genomes, 22% of infected cells to carry genomes with one mutation, 3.3% to carry two mutations, 0.33% to carry three mutations, and so on (Table 2).

Not all 10^{10} virions produced per day will infect other cells. Some will be defective; others, even if infectious, will be cleared. If, as suggested by Haase et al. [**11**], about 100 virions are produced per productively infected CD4$^+$ T cell, then at steady state only one of these 100 virions should go on to successfully infect another cell and produce a new generation of virions. (Recent, unpublished data from P. Bucy, University of Alabama, suggests that this estimate of 100 virions may be an order of magnitude too low. However, even if true it will not change the conclusions reached below.) If one productively infected cell led to the productive infection of more than one other cell, then the infection would not be in steady state and the number of virions would be increasing. If we assume that one out of a hundred virions infect another cell, then on average there are 10^8 new infections

Base changes	Probability of mutant	Number created/day	Number of possible mutants	Fraction of all possible mutants created/day
0	0.74	$7.4\text{x}10^7$	1	
1	0.22	$2.2\text{x}10^7$	$3.0\text{x}10^4$	1
2	0.033	$3.3\text{x}10^6$	$4.5\text{x}10^8$	$7.4\text{x}10^{-3}$
3	0.0033	$3.3\text{x}10^5$	$4.5\text{x}10^{12}$	$7.4\text{x}10^{-8}$

TABLE 2. Adapted from Table 3 of [**39**]. Assuming that the HIV-1 genome is approximately 10^4 base pairs long and that reverse transcriptase introduces errors into the HIV-1 genome at a rate of 3×10^{-5} per base pair per generation, the Table shows the probability of an infected cell containing a proviral HIV-1 genome with 0, 1, 2, or 3 mutations in it. Further, assuming 10^8 new cells are infected each day, the expected number of such mutants created each day within one infected individual is given. Lastly, we have also computed the fraction of all possible mutants with 1, 2, or 3 mutations that are created each day. If, for example, only one particular triple mutant gave rise to drug resistance then the probability of it being generated in a given patient on any given day from a wild-type virus is 7.4×10^{-8}.

per day. Hence, as shown in Table 2, we expect on average 0.22×10^8 mutants to be generated per day with single base changes. Because each of the 10^4 bases in HIV-1 could mutate to any of 3 other bases, there are a total of $3^i\binom{n}{i}$ possible sequences of length n with i mutations. Thus, with $n = 10^4$, there are 3×10^4 possible single base mutants, and essentially all of them would be generated each day. Consequently, resistance to drugs, such as 3TC, which only require a single base change is expected to occur rapidly and does [**44**]. In our initial studies with ritonavir [**16**], virus plasma levels fell to approximately 1% of their pretreatment values in 2 weeks. However, in 18 out of 20 patients the decrease in virus was not maintained for long periods and the viral load rebounded. Genetic analysis suggests that one of the key resistance mutations for the protease gene, a valine to alanine substitution at position 82, pre-existed [**8**], as predicted by these calculations. A similar pattern of rapid emergence of drug resistance to HIV-1 reverse transcriptase inhibitor nevirapine has also been seen [**13, 47**].

The chance of a mutant arising with a particular two base change, while substantially lower, is not so low that it can be neglected. The rate of generation of double mutants is 3.3×10^6/day or $\sim 0.74\%$/day of the possible two-base mutants (Table 2). Because a number of mutation combinations can lead to resistance, there is a reasonable chance that two-base mutants that could confer drug resistance pre-exist or will appear. In fact, in a population of size $1/7.4 \times 10^{-3} = 135$ one would expect every possible double mutant to be created on average each day. Although we have not modeled mutant survival, we would expect many of these mutants to survive and be able to grow in the presence of drug. Thus, in a large population, many recipients of drug will show resistance if only two mutations are required. Models which explicitly take into consideration the generation and growth of mutants have been developed [**3, 4, 32, 46, 48**] and will not be reviewed here.

For three-base change mutants, the situation is different. In single replication cycles, less than 10^{-7} of all possible three-base mutants are generated per day (Table 2). Thus, it is extraordinarily unlikely that any particular three-base change mutant will arise spontaneously. However, such mutants can be selected by sequential mutations if one or two-base mutants replicate.

8. Combination Therapy

The results of the last section suggest that therapy using a single drug is doomed to fail because of drug-resistance. Thus, in order to sustain a long-term response, models predict, and experience has borne out, that combination therapy is needed. Since both reverse transcriptase (RT) and protease inhibitors are available, most combination therapies employ both types of inhibitors in order to block the viral life cycle at two independent points.

In a clinical trial employing a protease inhibitor, nelfinavir, and two reverse transcriptase inhibitors, AZT and 3TC, novel dynamics were observed after therapy was initiated [**36**]. As with a single drug, the virus concentration in plasma fell dramatically for one to two weeks. However, under continued therapy, after this initial 'first phase' of decline, the virus continued to fall but at a significantly slower rate. When the virus concentration was plotted on a logarithmic scale, the two-phases of viral decline were both seen to be approximately exponential, as displayed in Figure 7.

The question then arose, why did the virus not continue to fall at the rapid first-phase rate? The antiretroviral effect of the combination therapy was potent in that the concentration of HIV-1 in plasma dropped below the standard detection threshold of 500 copies/ml by 8 weeks of treatment, and was found to be <25 copies/ml at week 16-20 using an ultrasensitive assay. Further, there was no evidence of emergence of drug-resistant virus during the 16-20 week study period, suggesting that the second phase of decay has a biological origin.

In the presence of both RT and protease inhibitors our basic model, with $T = constant = T_0$, takes the form

$$\begin{aligned} \frac{dT^\star}{dt} &= (1-\eta_{rt})kV_I T_0 - \delta T^\star, \\ \frac{dV_I}{dt} &= (1-\eta_p)N\delta T^\star - cV_I, \\ \frac{dV_{NI}}{dt} &= \eta_p N\delta T^\star - cV_{NI}, \end{aligned} \tag{42}$$

where η_{rt} and η_p are the effectiveness of the RT and protease inhibitors, respectively. If both the protease and RT inhibitors are 100% effective, then the model simplifies to

$$\begin{aligned} \frac{dT^\star}{dt} &= -\delta T^\star, \\ \frac{dV_I}{dt} &= -cV_I, \\ \frac{dV_{NI}}{dt} &= N\delta T^\star - cV_{NI}, \end{aligned} \tag{43}$$

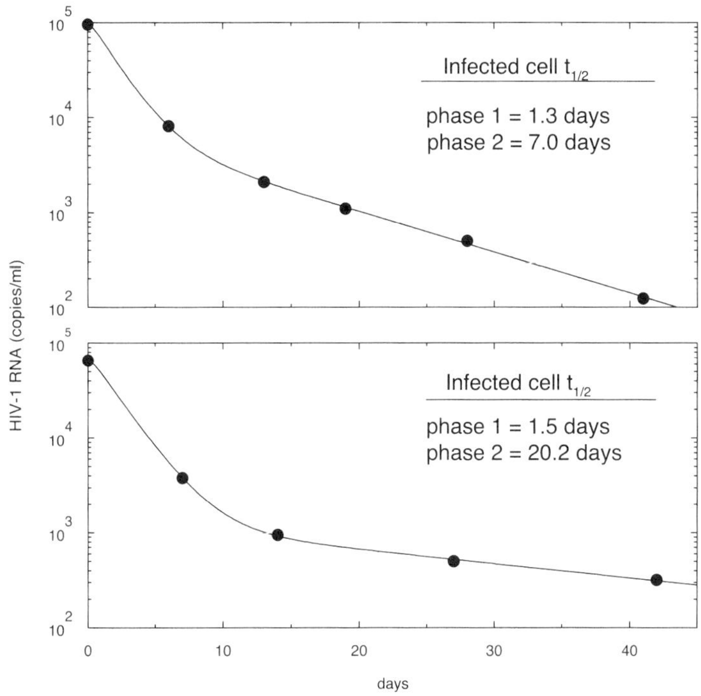

FIGURE 7. Examples of two-phase decline in plasma virus seen after the initiation of potent combination therapy. Data from two different patients is shown.

with solution

$$T^\star(t) = T^\star(0)e^{-\delta t} = \frac{kV_0T_0}{\delta}e^{-\delta t},$$

$$V_I(t) = V_0e^{-ct},$$

$$V_{NI}(t) = \frac{N\delta T^\star(0)}{c-\delta}\left[e^{-\delta t} - e^{-ct}\right].$$

If we assume that the patient was in quasi-steady state before treatment, then $T^\star(0) = kV_0T_0/\delta$ and $NkT_0 = c$, and we find

$$V(t) = V_I(t) + V_{NI}(t) = V_0\left[\frac{ce^{-\delta t} - \delta e^{-ct}}{c-\delta}\right].$$

Hence, one would expect that after a rapid transient of the order of $1/c$, the viral load would fall exponentially at a rate characterized by the constant δ. This fall corresponds to the first phase decline observed in the data.

Although there are many possible explanations for the second phase of plasma virus decline, we shall restrict our attention to those that we believe are the most plausible. First, if there were a source of virus other than infected cells, then the virus concentration, rather than falling continually, would ultimately come to a new steady state in which production by this source was in balance with clearance. If the source, rather than being constant, decayed slowly, then the virus concentration would exhibit a second phase decline whose rate reflected the rate of decay of the source.

In addition to CD4$^+$ T cells, other cells are known to be susceptible to HIV-1 infection. One such pool of cells are macrophages, large cells that reside in tissue and which can engulf various biological debris. A fraction of these cells are CD4$^+$, and in cell culture they become productively infected by HIV-1. Furthermore, HIV-1 infection in cell culture tends not to kill macrophages, and they are able to produce virus particles for weeks in culture. Another possibility is that the pool of T cells is heterogeneous, with a subpopulation of T cells upon infection producing fewer virions and living longer. Incorporating a second population of cells, M, susceptible to infection with rate constant k_M, we obtain the following, which we call the long-lived infected cell model:

$$
\begin{aligned}
\frac{dT}{dt} &= s + pT(1 - \frac{T}{T_{max}}) - d_T T - kVT, \\
\frac{dT^\star}{dt} &= kVT - \delta T^\star, \\
\frac{dM}{dt} &= s_M - d_M M - k_M V M, \\
\frac{dM^\star}{dt} &= k_M V M - \mu_M M^\star, \\
\frac{dV}{dt} &= N\delta T^\star + p_M M^\star - cV.
\end{aligned} \tag{44}
$$

Here $M^\star$ denotes the concentration of productively infected long-lived cells, which are assumed to produce virus continually at rate p_M per cell and to die at rate μ_M per cell. Long-lived cells are assumed to be created by a constant source at rate s_M and to die with rate constant d_M.

Before treatment is initiated we shall assume that patients are in quasi-steady state, so that at $t = 0$, $dV/dt = 0$, $dT^\star/dt = 0$, and $dM^*/dt = 0$, or

$$
T_0^\star = kV_0T_0/\delta, \quad M_0^* = k_M V_0 M_0/\mu_M, \quad N\delta T_0^\star + p_M M_0^* = cV_0 \ . \tag{45}
$$

If both the protease and reverse transcriptase inhibitors are 100% effective, then for $t > 0$, $k = k_M = 0$, and all virus that is produced is non-infectious. Hence, after therapy is initiated

$$
\begin{aligned}
\frac{dT}{dt} &= s + pT(1 - \frac{T}{T_{max}}) - d_T T, \\
\frac{dT^\star}{dt} &= -\delta T^\star, \\
\frac{dM}{dt} &= s_M - d_M M, \\
\frac{dM^\star}{dt} &= -\mu_M M^\star, \\
\frac{dV_I}{dt} &= -cV_I, \\
\frac{dV_{NI}}{dt} &= N\delta T^\star + p_M M^\star - cV_{NI}.
\end{aligned} \tag{46}
$$

The equations for T and M decouple from the equations determining viral dynamics, and hence the assumption that T or M remains constant after therapy need not be made. Because the viral dynamic equations are linear, the following solution can be obtained in a straightforward manner.

$$\begin{aligned} T^{\star}(t) &= T_0^{\star} e^{-\delta t} , \\ M^{*}(t) &= M_0^{*} e^{-\mu_M t} , \\ V_I(t) &= V_0 e^{-ct} , \\ V_{NI}(t) &= \frac{N\delta T_0^{\star}}{c-\delta}[e^{-\delta t} - e^{-ct}] + \frac{pM_0^{*}}{c-\mu_M}[e^{-\mu_M t} - e^{-ct}] . \end{aligned} \tag{47}$$

Using the quasi-steady state constraints (45), we find

$$V_{NI}(t) = V_0\Big[\frac{NkT_0}{c-\delta}(e^{-\delta t} - e^{-ct}) + \frac{c-NkT_0}{c-\mu_M}(e^{-\mu_M t} - e^{-ct})\Big] ,$$

and hence

$$V(t) = V_0\Big[\Big(1 - \frac{NkT_0}{c-\delta} - \frac{c-NkT_0}{c-\mu_M}\Big) e^{-ct} + \frac{NkT_0}{c-\delta} e^{-\delta t} + \frac{c-NkT_0}{c-\mu_M} e^{-\mu_M t}\Big] . \tag{48}$$

Using nonlinear least squares regression we fitted the patient data obtained in this combination therapy clinical trial to the long-lived infected cell model and estimated three parameters: δ, μ_M, and the composite parameter, NkT_0 [**36**]. Virion clearance occurred too rapidly to estimate c from the data obtained in this study. Thus c was held constant at the mean value, c =3 day^{-1}, determined in [**40**]. The parameter μ_M, which determines the lifespan of the long-lived infected cells, ranged from 0.03 to 0.12 day^{-1}, with a mean value of 0.07$\pm$ 0.04 day^{-1}. The half-life of the cell population responsible for the second phase of viral decay, $t_{1/2} = ln2/\mu_M$ had a mean value of 13.3 days, with a standard deviation of 7.9 days. The theoretical curves generated from (48), using the best-fit parameter values gave an excellent fit to the data for all patients (see Figure 7 for examples). The values of δ ranged from 0.36 to 0.81 day^{-1}, with a mean of 0.59 $\pm$ 0.15 day^{-1}. The corresponding $t_{1/2}$ values of the short-lived infected cells ranged from 0.86 to 1.92 days, with a mean of 1.25 $\pm$ 0.34 days. This mean value was somewhat smaller than our previous estimate of 1.6 days [**40**], possibly due to the greater effectiveness of the triple drug combination than the monotherapy used in [**40**].

While the long-lived cell model fit the patient data, it is not the only reasonable biological model [**36**]. We will show a model that includes delay dynamics in the following section and we refer the reader to other possible models of long term infection in [**39**]

9. Delay Models

Delay is inherent in the immune system's response to disease. There is, for example, delays during cell division, delays in viral production from the time a cell is first infected and delays in activation of some medications from a prodrug to an active form. With the discovery of HIV-1's rapid dynamics it is relevant to examine the effect delays have on model dynamics. To accomplish this we examine aspects of the viral life cycle that have not been addressed to this point, such as the intracellular delays associated with the cell-virus interaction. In HIV-1 infection, the virus life cycle plays a crucial role in disease progression. The binding of a viral particle to a receptor on a target cell initiates a cascade of events that can ultimately lead to the target cell becoming productively infected, i.e. producing new virus. The previous models (11), (18), (42) and (44) assumed this process to occur instantaneously. In other words, they assumed that as soon as HIV-1 infects a cell that cell begins producing virus. However, in reality there is a time delay between initial viral entry into a cell and subsequent viral production. This delay needs to be taken into account to accurately determine the half life of free virus and the rate of loss of T cells once they begin producing virus from drug perturbation experiments [**14, 25, 28**]. However, if the drug is assumed to be completely efficacious, the time delay does not affect the estimated rate of decay of virus producing T cells [**29, 30**].

The dynamics of HIV-1 decay observed in patients after potent antiretroviral therapy is initiated has distinct phases. For the models considered here we only use data from patients whose viral load is stationary before initiation of therapy, i.e. patients in the asymptomatic phase of viral infection. After initiation of drug therapy there is a period that lasts anywhere from 6 hours to a few days where little change is noticed in the virus concentration [**40**]. During this period the viral decay curve is said to have a "shoulder". The delay or shoulder is believed to be due in part to drug dynamics such as a pharmacological delay, i.e., a delay associated with the transport and intracellular processing of drug. Second, if protease inhibitors are given, infected cells continue to produce virus, but the virus is noninfectious. Thus, the virus concentration does not fall until there is a decrease in the number of virus producing cells. Third, if a reverse transcriptase inhibitor is given, cells that have already had the virus enter and its RNA reverse transcribed, will continue to go through the set of intracellular events that slowly lead to virus production [**14, 26, 40**]. Fourth, there is a delay between the time a cell is infected and its beginning production of virus particles, i.e., what we call an intracellular delay.

After the shoulder the viral decline is at first very rapid (first phase) and then around two weeks after therapy initiation the rate of decline slows (second phase) [**36**]. Perelson *et al.* [**40**] showed that the rate of plasma virus decline during the first, rapid, phase could provide an estimate for δ, the decay rate of productively infected T cells. However, the effectiveness of the drug [**3, 7, 28, 48**] and the intracellular delay [**26, 28, 29**] also contribute to the rate of viral plasma decline. We found an analytical expression, $\lambda_d \sim -\delta\eta_p C(\eta_p, \delta, \tau)$, where λ_d is the controlling eigenvalue for $t > 1$ day and $C(\eta_p, \delta, \tau) = \frac{1}{1+(1-\eta_p)\delta\tau}$, which highlights the effects of combining intracellular delays with less than perfect drug therapy on the estimated value of δ [**29, 30**].

9.1. Fixed delay models. The first study of HIV-1 patient data using models with time delays was done in 1996 by Herz *et al.* [**14**]. To incorporate the intracellular phase of the viral life cycle, they assumed that the production of viral particles by T cells occur τ days after initial infection. The model equations in our notation are

$$
\begin{aligned}
\frac{dT}{dt} &= \lambda - d_T T(t) - k(t)T(t)V(t), \\
\frac{dT^\star}{dt} &= k(t-\tau)T(t-\tau)V(t-\tau)e^{-m\tau} - \delta T^\star(t), \qquad (49)\\
\frac{dV}{dt} &= N\delta T^\star(t) - cV(t),
\end{aligned}
$$

where m is the death rate of cells between initial infection and the beginning of viral production τ time units later. Assuming death is a first order process, the probability a cell infected at time t is alive at time $t+\tau$ is then $e^{-m\tau}$.

Assuming monotherapy with an RT inhibitor that is completely efficacious, i.e., $k(t)=0$, $t>0$ and that the system is initially in steady state with $V(0)=V_0=\frac{N\delta T^\star(0)}{c}$, $T^\star(0)=T_0^\star=\frac{\lambda}{\delta}e^{-m\tau}-\frac{d_T c}{Nk\delta}$, and $T(0)=T_0=\frac{c}{kN}e^{m\tau}$, one can solve (49) for $0\le t<\tau$ to obtain

$$\begin{aligned} T(t) &= \lambda/d_T + (T_0-\lambda/d_T)e^{-d_T t},\\ T^\star(t) &= T_0^\star,\\ V(t) &= V_0, \end{aligned} \tag{50}$$

and for $t\ge\tau$

$$\begin{aligned} T(t) &= \lambda/d_T + (T_0-\lambda/d_T)e^{-d_T t},\\ T^\star(t) &= T_0^\star e^{-\delta(t-\tau)},\\ V(t) &= \frac{V_0}{c-\delta}\left[ce^{-\delta(t-\tau)}-\delta e^{-c(t-\tau)}\right] \end{aligned} \tag{51}$$

Notice the solution for $T(t)$ is independent of τ and that the solutions for $T^\star(t)$ and $V(t)$ are independent of $T(t)$.

If one considers monotherapy with a protease inhibitor and assumes $T(t)=T_0$ for all t, then one has the following model

$$\begin{aligned} \frac{dT^\star}{dt} &= kT_0V_I(t-\tau)e^{-m\tau}-\delta T^\star,\\ \frac{dV_I}{dt} &= (1-\eta_p)N\delta T^\star - cV_I,\\ \frac{dV_{NI}}{dt} &= \eta_p N\delta T^\star - cV_{NI}. \end{aligned} \tag{52}$$

For 100% effective therapy a solution is easily attained as (see [**14**])

(53)

$$V(t)=V_0,\qquad 0\le t\le\tau,$$

$$V(t)=V_0e^{-c(t-\tau)}+\frac{cV_0}{c-\delta}\left[\frac{c}{c-\delta}\left[e^{-\delta(t-\tau)}-e^{-c(t-\tau)}\right]-\delta(t-\tau)e^{-c(t-\tau)}\right],\qquad t>\tau$$

where $V(t)=V_{NI}(t)+V_I(t)$. Without an intracellular delay, $\tau=0$, (53) reduces to (22).

While (49) and (52) were the first models to include time delays, the models assumed monotherapy with treatment being completely effective. In [**29**], we relaxed this assumption of 100% effective drug therapy in (52).

The equation for $V_{NI}(t)$ can be easily solved once the solution for the first two equations is known. In order to compare the stability of the model (18) without delay and (52), we analyzed the characteristic equation of (52),

$$(\lambda^2+(\delta+c)\lambda+\delta c-\psi e^{-\lambda\tau})(\lambda+c)=0 \tag{54}$$

where $\psi=(1-\eta_p)\delta NkT_0e^{-m\tau}$. Assuming that the system is in quasi-steady state before drug therapy, i.e., $c=NkT_0$, and that the death during transition from infected to productively infected is negligible, i.e., $m<<1$, we simplified ψ to

$$\psi=(1-\eta_p)\delta c. \tag{55}$$

It is easy to see that one of the eigenvalues of (54) is $-c$. To determine the remaining two eigenvalues, which determines if the time delay introduces instabilities, we need to solve

$$H \equiv \lambda^2 + (\delta + c)\lambda + \delta c - \psi e^{-\lambda\tau} = 0. \tag{56}$$

Let $\sigma(\tau)$ be the supremum of the real parts of the eigenvalues satisfying (56). It is well known [**12**] that $\sigma(\tau) < 0$ implies the zero solution of (52) is uniformly asymptotically stable. The characteristic equation is continuous in τ since the supremum $\sigma(\tau)$ is attained at some specific value of λ satisfying (56). Hence it follows that $\sigma(\tau)$, given $\partial H/\partial\lambda \neq 0$, is continuous in τ. This result allows one to study the eigenvalue equation to determine stability. We first show the conditions for which (56) has all eigenvalues negative for the case $\tau = 0$, i.e., all eigenvalues are in the left half of the complex plane. Once this is determined, we then consider the case of $\tau > 0$ to see if any of the previous eigenvalues that have negative real part cross the imaginary axis. If not then the time delay does not bifurcate the system and we can compare the results between the non-delay and the time-delay models.

For (52) we showed stability was maintained with the following result (see [**29**]). Let $H = \lambda^2 + (\delta + c)\lambda + \delta c - \psi e^{-\lambda\tau}$. Set $\lambda = \mu + i\nu$ and let $H(\mu) = \mu^2 + (\delta + c)\mu + \delta c - \psi e^{-\mu\tau}$.

LEMMA 1. *If $H(\mu) > 0$ then there is no $\nu \in \mathbb{R}$ such that $H(\mu + i\nu) = 0$. Moreover, if $H(\mu) = 0$ then $H(\mu + i\nu) = 0$ if and only if $\nu = 0$.*

Proof by contradiction: Assume that $H(\mu + i\nu) = 0$. Then $H(\mu + i\nu)$ can be rewritten in terms of its real and imaginary parts as $\mu^2 - \nu^2 + \mu(\delta + c) + \delta c = \psi e^{-\mu\tau} \cos\nu\tau$ and $2\mu\nu + \nu(\delta + c) = -\psi e^{-\mu\tau} \sin\nu\tau$. Taking the squares of the absolute values of both equations and adding together gives $[\nu^2 + (\mu + c)^2][\nu^2 + (\mu + \delta)^2] = \psi^2 e^{-2\mu\tau}$ which implies that $(\mu + c)^2(\mu + \delta)^2 - \psi^2 e^{-2\mu\tau} = -\nu^4 - \nu^2[(\mu + c)^2 + (\mu + \delta)^2]$. The left hand side of this equation equals $H(\mu)[(\mu + c)(\mu + \delta) + \psi e^{-\mu\tau}] = H(\mu)[H(\mu) + 2\psi e^{-\mu\tau}]$, which is always positive given $H(\mu) > 0$. The right hand side is always negative leading to a contradiction. Therefore, $H(\mu + i\nu) \neq 0$ given $H(\mu) > 0$. The second part of the proof is easily seen.

Note that $H(\mu)$ can be rewritten as $H(\mu) = \mu^2 + (\delta + c)\mu + \delta c(1 - (1 - \eta_p)e^{-\mu\tau})$. Since $\delta > 0$, $c > 0$, and $0 \leq \eta_p \leq 1$, $H(\mu) > 0$ for $\mu > 0$. Therefore by Lemma 1 there are no eigenvalues with non-negative real part. Moreover, a special case of the above lemma shows that if $H(0) > 0$, then there is no $\nu \in \mathbb{R}$ such that $H(i\nu) = 0$. Hence there are no purely imaginary roots.

With the previous result we then showed that the model (52) had only one eigenvalue in a vertical strip of the complex plane bounded by $(-3, 0)$. This was significant in that it gave us justification for the following result which determines an analytical approximation for this eigenvalue, which we call the dominant eigenvalue λ_d. Using asymptotic methods we determined an analytical expression for λ_d that contains the scaling term $C(\eta_p, \delta, \tau)$ from (52). We did this by first expanding the $e^{-\lambda\tau}$ term in (56) in a Taylor series to get

$$e^{-\lambda\tau} \sim 1 - \lambda\tau + \frac{\lambda^2\tau^2}{2} + O(\lambda^3\tau^3). \tag{57}$$

This expression is valid for $|\lambda\tau| << 1$ and we provide justification for this in [**29, 30**]. Substituting this into (56) we find

$$\lambda^2 + \lambda(\delta + c) + \delta c - \psi(1 - \lambda\tau + \frac{(\lambda\tau)^2}{2} + 0(\lambda^3\tau^3)) = 0 \tag{58}$$

and solving for λ gives

$$\lambda = \frac{1}{2(1 - \frac{\psi\tau^2}{2})}[-(\delta + c + \psi\tau) \pm (\delta + c + \psi\tau)\sqrt{1 - \frac{4(\delta c - \psi)(1 - \frac{\psi\tau^2}{2})}{(\delta + c + \psi\tau)^2}}], \tag{59}$$

where $\psi = (1 - \eta_p)\delta c$. The square root term can be approximated by the first two terms of its binomial expansion, since for the parameter values of interest $|\frac{4(\delta c - \psi)(1 - \frac{\psi\tau^2}{2})}{(\delta + c + \psi\tau)^2}| < 1$. Using this and taking the positive term we get

$$\lambda_d \sim -\frac{\delta c - \psi}{\delta + c + \psi\tau}. \tag{60}$$

Assuming $c >> \delta$ and using the definition of ψ we find

$$\lambda_d \sim -\frac{\delta\eta_p}{1 + (1 - \eta_p)\delta\tau} = -\delta\eta_p C(\tau, \eta_p, \delta), \tag{61}$$

where $C(\tau, \eta_p, \delta) = \frac{1}{1 + (1 - \eta_p)\delta\tau} < 1$ is the correction introduced by incorporating a delay. Notice the rate of decay, δ, is proportional to the product of the rate constant for death of virus producing T cells, δ, and the efficacy of the protease inhibitor, η_p, but now this product is reduced by the delay since as τ gets bigger the value of δ gets smaller. This result shows why earlier work [**14, 26**] found no change in the predicted value for δ in delay models; by assuming a perfect drug $\eta_p = 1$, the delay effect canceled out.

We illustrate the effects of delay in Figure 8 and the effects of a less than perfect antiviral therapy in Figure 9. When $\eta_p \neq 1$, the delay reduces the long term rate of decline of the viral load and hence any estimate of δ that is made which does not take the delay into account will be flawed. The effect, however, is not large. For example, if $\tau = 1$ day, a typical value, and $\eta_p = 0.8$ then the correction due to the delay, $C(\tau, \eta_p, \delta) = \frac{1}{1+0.2\delta}$. For $\delta = 0.5$ day^{-1} [**40**], $C = 0.91$. If the drug is only 50% efficacious then $C = 0.8$. Hence, if the intracellular delay is approximately constant, as assumed here, and if the effects of drug efficacy are properly taken into account, then including the delay will change the estimate of δ from $\frac{\text{slope}}{\eta_p}$ to $\frac{\text{slope}}{\eta_p C}$, (where slope is the slope of the first phase of viral decay estimated from data) which for the cases considered above yields at most a 25% increase in δ.

9.1.1. *Fits of discrete time delay model to clinical data.* Using non-linear least squares regression we fit numerical solutions of (52) to the decline in HIV-1 RNA in patients who received the protease inhibitor ritonavir [**40**]. For $\tau = 0$ and $\eta_p = 1.0$ the average estimates for δ and c were 0.47 days^{-1} and 3.1 days^{-1}, respectively, similar to that given in [**40**]. As expected from analyses in [**40**], fitting the data with lower values of η_p, resulted in higher estimates for δ. The average change in δ was 0.49 days^{-1} to 0.69 days^{-1} as the theoretical efficacy of the protease inhibitor was reduced from 100% to 85% (see [**28**] for results with drug therapies less than 85 %.) We found, as discussed in [**14, 26**], that estimates for c given in [**40**] are minimal estimates since the model does not account for the intracellular phase of the viral life cycle. Combining the intracellular delay with less than perfect antiviral

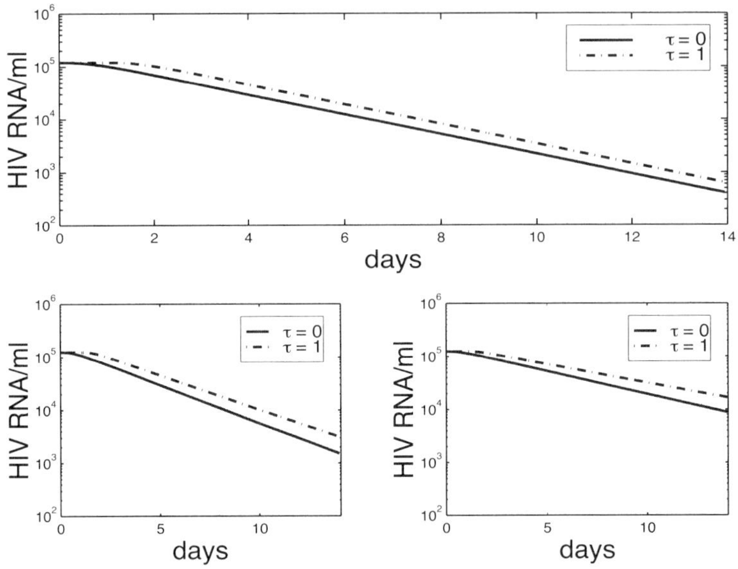

FIGURE 8. The top plot shows the numerical solution for viral decay when a perfect drug is used. Notice the slopes of the decay with and without a delay are parallel. The bottom two graphs compare the viral decay with and without a delay assuming a less than perfect drug; with $\eta_p = 0.8$ in the left panel and $\eta_p = 0.5$ in the right panel. Both cases show the slope is affected by the delay.

therapy leads to significant increases in the previously estimated value for δ (see Table 3 and Table 4).

9.2. Distributed time delay models. Mittler *et al.* [**26**] examined a linear model similar to (42) but which included a continuous delay. The model was

$$\begin{aligned} \frac{dT^\star}{dt} &= \int_0^\infty kT_0 f(t')V_I(t-t')e^{-mt'}dt' - \delta T^\star(t), \\ \frac{dV_I}{dt} &= [1-h(t)](1-\eta_p)pT^\star(t) - cV_I(t), \\ \frac{dV_{NI}}{dt} &= h(t)\eta_p pT^\star(t) - cV_{NI}(t). \end{aligned} \tag{62}$$

The first equation of (62) includes an integral term that examines the effects of assuming a continuous delay on the viral dynamics, t' represents the delay variable, and $f(t')$ was chosen to be a gamma function where

$$f(t') = g_{n,b}(t') \equiv \frac{(t'^{(n-1)}}{(n-1)!b^n}e^{-t'/b}, \tag{63}$$

and $h(t)$ is a Heaviside function that allowed for varying the start time of the drug therapy. As was done for (49), this model was analyzed assuming that the drug therapy was completely effective, i.e., $\eta_p = 1$. Using this assumption it is possible

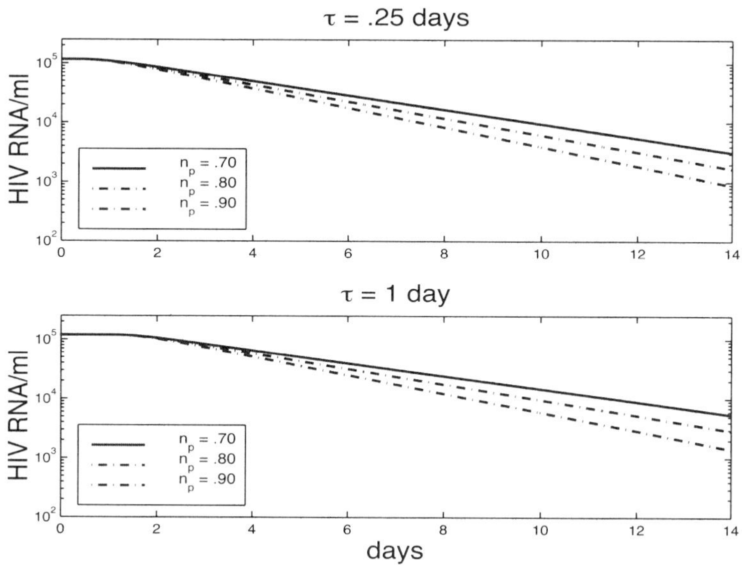

FIGURE 9. The difference in the predicted decay of virus with varying drug effectiveness. Notice the slope of the curve from day 2 to day 7 depends on n_p, which shows the effect that less than perfect antiviral therapy has on the estimated decay rates.

Estimates for (c, δ) with no delay

	Efficacy (%)			
Patient	**100**	**95**	**90**	**85**
102	(4.4, .26)	(4.4, .27)	(4.4, .29)	(4.2, .31)
103	(2.4, .65)	(2.4, .69)	(2.4, .74)	(2.4, .79)
104	(3.7, .46)	(3.6, .49)	(3.7, .52)	(3.5, .56)
105	(2.0, .47)	(2.0, .51)	(1.9, .55)	(2.0, .59)
107	(2.8, .49)	(2.8, .52)	(2.8, .56)	(2.8, .60)
Average	**(3.06, .47)***	**(3.04, .50)**	**(3.04, .53)**	**(2.98, 0.57)**
Stdev	**(.87, .12)**	**(.86, .13)**	**(.86, .14)**	**(.79, .15)**

TABLE 3. Estimates for c and δ from a model without an intracellular delay. Estimates were obtained from non-linear least squares fits of numerical solutions to (52) with $\tau = 0$ to data from [**40**]. In each case, the value for the uninfected target cells was assumed equal to the baseline concentration of uninfected CD4$^+$ T cells given in [**40**].

*estimates for the 100% efficacy case differ slightly from those in Perelson *et.al.* (1996) because we did not use a jack-knife procedure and because we assume a pharmacological delay of 2 hours for each patient, whereas Perelson *et. al.* estimated the pharmacological delay (which they constrained to be between 2-6 hours from the data).

Estimates for (c, δ) with delay

	Efficacy (%)				τ
Patient	**100**	**95**	**90**	**85**	**(days)**
102	(11.2, .27)	(11.2, .30)	(11.1, .31)	(11.1, .33)	0.50
103	(4.4, .73)	(4.1, .83)	(3.6, .98)	(3.2, 1.21)	0.99
104	(5.4, .48)	(5.4, .54)	(5.4, .56)	(5.4, .61)	0.50
105	(4.8, .49)	(4.7, .55)	(4.6, .61)	(4.6, .66)	0.72
107	(4.9, .49)	(4.9, .55)	(4.9, .60)	(4.8, .65)	0.51
Average	**(6.14, .49)**	**(6.06, .55)**	**(5.92, .61)**	**(5.82, .69)**	**0.64**
Stdev	**(2.54, .15)**	**(2.58, .17)**	**(2.66, .21)**	**(2.71, .28)**	**0.23**

TABLE 4. Estimates for (c, δ) and τ from a model with an intracellular delay. Estimates obtained in same manner as in Table 3, but allowing τ to vary between 0.5 and 1.5 days. For all patients allowing a delay provided a better fit to the data than the non-delay model.

to solve (62) to get

$$V_I(t) = V_I(0)e^{-ct} \tag{64}$$

and

$$\begin{aligned} V_{NI}(t) = V_I(0)\{&\left[G_n ct - (\frac{c}{\delta - c})H_n + (\frac{c\bar{b}}{1-c\bar{b}})\sum_{k=0}^{n-1}\frac{1 - G_{n,k} + H_{n,k}}{(1-c\bar{b})^k}\right]e^{-ct} \\ &+(\frac{c}{\delta - c})H_n e^{-\delta t} - (\frac{c\bar{b}}{1-c\bar{b}})\sum_{k=0}^{n-1}\frac{1 - G_{n,k} + H_{n,k}}{(1-c\bar{b})^k}\sum_{l=0}^{k}\frac{(1-c\bar{b})^l t^l}{l!\tau^l}e^{-t/\bar{b}}\} \end{aligned} \tag{65}$$

where $\bar{b} = \frac{1}{1+mb}$ comes from rescaling (62) to remove the $e^{-mt'}$ term and

$$\begin{aligned} G_i &\equiv (\frac{\delta}{\delta - c})(1 - c\bar{b})^{-i}, \\ H_i &\equiv (\frac{\delta}{\delta - c})(1 - \delta\bar{b})^{-i}. \end{aligned} \tag{66}$$

The total viral concentration is then given by $V(t) = V_I(t) + V_{NI}(t)$.

To determine the effect of neglecting the time delay on parameter estimations, Mittler *et al.* [**26**] generated simulated data by solving the delay model with different but biologically reasonable delay kernels. They then fit that data to a model without delay. They found that the previously estimated values for c [**40**] were underestimates. The underestimation could be approximated analytically if one fits a model without delay to viral decay curves in which the underlying delays are gamma distributed. One condition for a good match between models is that their solutions approach each other in the limit as $t \to \infty$. To match two gamma kernels of order n and n', they must satisfy $H_n c/(\delta - c) = H_{n'} c'(\delta - c')$ or

$$\frac{c'}{\delta} = \frac{c}{c + (\delta - c)\sqrt{\frac{(1-\delta\bar{b})^n}{(1-\delta\bar{b})^{n'}}}} \tag{67}$$

for $\delta < c$ and $\delta < c'$. Hence the clearance rate c' of the non-delayed model, i.e. $\bar{b} = 0$, is related to the parameters of a gamma distributed delay model by

$$c' = \frac{\delta c}{c + (\delta - c)\sqrt{(1 - \delta \bar{b})^n}} \tag{68}$$

where c and δ are the previously estimated parameters. Their analysis however did not find any changes in the previously estimated value for δ due to their assumption of a perfect drug.

10. Discussion

We have tried to show how mathematical modeling has impacted our understanding of HIV-1 pathogenesis. Before modeling was brought to bear in a serious manner, AIDS was thought to be a slow disease in which treatment could be delayed until symptoms appeared and patients were not monitored very aggressively. In the large multicenter AIDS cohort studies aimed at monitoring the natural history of the disease, blood typically was drawn every six months. There was a poor understanding of the biological processes that were responsible for the observed levels of virus in the blood and the rapidity at which the virus became drug resistant. Modeling, coupled with advances in technology, has changed all of this. In section 4, we showed how an extremely simple model involving a single linear ordinary differential equation, when applied to the interpretation of clinical data obtained in a phase I/II drug trial, gave the first quantitative estimate of how rapidly HIV-1 was being produced and cleared in an infected person. While the mathematics involved was trivial, the application of mathematics in this manner was novel, and set off what has been described as a revolution in thinking about HIV-1. The papers by Ho et al. [**16**] and Wei et al. [**47**] that were published in the same issue of Nature and which both reported an approximately 2 day half-life for HIV-1 in plasma were the most highly cited scientific papers published in 1995.

A more complex model presented in section 5 that incorporated both virus and infected cells, when compared with data collected to test the model, showed that the two day viral half-life mainly reflected the lifetime of infected cells that produced virus. Thus, the first estimate of the lifetime of a productively infected cell in vivo was obtained, which allowed us to think in a more quantitative manner about the issue of CD4$^+$ T cell depletion, the hallmark of AIDS. Further, the estimate of the rate of clearance of free viral particles was improved and the half-life of free virus particles in plasma was estimated at 6 hours or less. Introducing delay models has helped further refine the estimate for c [**25**]. Because the level of virus is maintained at steady-state, the rate of viral clearance can be used to determine how rapidly HIV-1 is produced. Doing these calculations led to the conclusion that in an average HIV-infected person at least 10^{10} virus particles are produced and released into bodily fluids each day. As explained in section 6.4, these estimates can be used to calculate a viral generation time of approximately 1.8 days. This implies that in an infected person HIV-1 can go through about 200 replication cycles per year, with the possibility of mutating at each replication. Thus, the rapid evolution of HIV-1 can easily be understood. The practical implication of this was that therapy with a single drug in which a few mutations were all that were required for resistance to arise could be shown to be a poor strategy. This helped herald in the current strategy of combination therapy and the approach of treating HIV-1 infected persons as soon after diagnosis as possible.

Using more complex models, involving multiple cell populations, has allowed further interpretation of clinical data obtained from patients on combination therapy in which the virus concentration in plasma has a two-phase decline. This work, described in section 8, has also had important practical consequences. By extrapolating in a rigorous way the data obtained from patients responding well to combination therapy we were able to estimate how long therapy would need to be given to clear the cells responsible for producing the observed levels of virus. While bringing home the lesson that therapy would need to continue for years after free virus became undetectable in blood, the work also began the process of quantifying both the level and role of latently infected and long-lived infected cell populations in HIV-1 infection.

Section 9 began to look at even more complex models which account for intracellular delays, less than perfect drug therapy and non-linearities. Including a delay also changes the estimated lifespan of a productively infected cell. In previous work without delays the average lifespan of a productively infected cell was calculated as $1/\delta$. We have shown that taking into account the period where no virus is produced, i.e. eclipse phase, of the viral life cycle where the length of the eclipse phase is represented by the delay τ, the average lifespan of cell from infection to death changes from $1/\delta$ to $\tau + 1/\delta$. We have been able to estimate the value of τ to be about 0.64 days and hence with a new estimate for δ of 0.69 days, we find the new life span of a productively infected T cell to be $0.64 + 1/0.69 = 2.09$ [**28**]. Thus, to a first approximation T cells infected with HIV-1 might live an average of 2 days rather than the average of 1-2 days previously reported [**40, 47, 16**]. Currently, work is in progress examining delays in drug activation and modeling the efficacy of the drug as a dynamically varying parameter. Also, there is a need to explore the effects of different types of delays such as distributed delays or multiple discrete delays; some of which is already in progress. The length of the delay is not directly measurable in vivo. Thus, additional in vitro data would prove valuable in expanding our knowledge of the delay characteristics.

Throughout this paper we have analyzed situations in which models predict that under sufficiently intense therapy, the virus and infected cell concentrations will go to zero. The question then arises, is viral eradication a realistic expectation? At the moment, we do not know. Clearly, the models that we have analyzed are only simplified caricatures of reality. They have not included the spatial and compartmental aspects of the body, and have implicitly assumed that drug is available everywhere in the body at a constant effectiveness. Not all drugs penetrate the blood-brain barrier effectively, and thus drug concentrations in the brain and central nervous system tend to be lower than in the circulation. Also, the cells of the immune system have limited access to these sites, and hence the brain and other sites with poor drug penetrance, such as the testes, may act as sanctuaries for the virus. Not surprisingly, in models with drug sanctuaries it is easier to generate drug resistance [**19**], and hence ongoing viral replication may remain a problem until a new generation of drugs become available.

The models have only dealt with the major targets of HIV-1 infection, $CD4^+$ T cells and macrophages. However, other cells may become HIV-infected and exhibit different kinetics. We have only dealt with infection of cells by free virus and the death of cells due to viral infection. Direct cell-to-cell transmission of virus has been reported in cell culture, as has death of cells due to effects other than direct viral killing. All of these features may play a role in the long-term behavior of HIV-1 in vivo.

Another large omission in our models has been the immune response. Processes such as the death of virally infected cells and/or the clearance of free viral particles may have an immune component. Furthermore, viruses can evolve to become more pathogenic and drug resistant. Thus, parameters in our models, rather than being constant, as we have assumed, may vary in time and depend in some complex way on events occurring in the host. Many of us hope that drugs will not need to be given until every last viral particle and infected cell is eliminated, but rather that as the amount of virus is reduced by antiretroviral drugs, recovery of the immune system will occur, and the host defense system will ultimately be able to control or clear any remaining virus. Both models and experiment are needed to examine this possibility.

Potent combination therapy has been available for 3 to 4 years. Analysis of lymphocytes from blood and tissue biopsies have shown that small numbers of latently infected cells can be found in patients treated for as long as 2.5 years [**10, 50**]. Whether the immune system can control the remaining infection is not known, and thus cessation of antiretroviral treatment is not recommended. The rate at which latently infected cells are decaying, and particularly those carrying proviruses that can generate infectious particles on stimulation, is also not yet known although there are some suggestions that they may have a half-life of ~ 3 - 5 months [**15**]. Making estimates of the pool size of latently infected cells one can surmise that 5 to 7 years of continuous therapy may be required to achieve eradication of latently infected $CD4^+$ T cells [**15**]. This is an unacceptable solution due to the complexity of complying with current drug regimes, drug side effects and cost. Alternative approaches are needed and are being examined [**15**].

Models, such as the ones discussed here, have been developed and tested on data obtained over relatively short periods: days, weeks, and in some cases months. Even though the models appear accurate on these time scales, they should not be used to predict the long-term events in individual patients. They have provided much insight into the biological events underlying the disease process, and have helped guide treatment strategies. Eradicating HIV-1 from an infected patient or helping the body control the infection still remains our goal. We believe that modeling will continue to play an important role in attaining this goal.

Acknowledgements

The work described here would not have been possible without extensive collaboration with David D. Ho, head of the Aaron Diamond AIDS Research Center, Rockefeller University, New York. Dr. Martin Markowitz, Aaron Diamond AIDS Research Center, was responsible for the clinical aspects of the studies reported here. Avidan Neumann of Bar-Ilan University, Israel and Paulina Essunger contributed to the original analysis of the patient data.

Portions of this work were performed under the auspices of the U.S. Department of Energy. This work was supported by NIH grants RR06555 and AI40387.

References

[1] R.M. Anderson. *Population Dynamics of Infectious Diseases*. Chapman and Hall, 1982.

[2] R.M. Anderson and R.M. May. *Infectious diseases of humans:Dynamics and control.* Oxford University Press, 1991.

[3] S. Bonhoeffer, R.M. May, G.M. Shaw, and M.A. Nowak. Virus dynamics and drug therapy. *Proc. Natl. Acad. Sci. USA*, 94:6971–6976, 1997.

[4] S. Bonhoeffer and M.A. Nowak. Pre-existence and emergence of drug resistance in HIV-1 infection. *Proc. R. Soc. London*, 264:631–37, 1997.

[5] T.W. Chun, L. Carruth, D. Finzi, X. Shen, J. A. DiGiuseppe, H. Taylor, M. Hermankova, K. Chadwick, J. Margolick, T.C. Quinn, Y.H. Kuo, R. Brookmeyer, M.A. Zeiger, P. Barditch-Crovo, and R.F. Siliciano. Quantification of latent tissue reservoirs and total body viral load in HIV-1 infection. *Nature*, 387:183–88, 1997.

[6] R. J. De Boer and A. S. Perelson. Target cell limited and immune control models of HIV infection: a comparison. *J. Theoret. Biol.*, 190:201–4, 1998.

[7] A.A. Ding and H. Wu. Relationship between antiviral treatment effects and biphasic decay rates in modeling HIV dynamics. *Math. Biosci.*, 160:63–82, 1999.

[8] P. S. Eastman, J. E. Mittler, R. Kelso, C. Gee, E. Boyer, J. Kolberg, M. Urdea, J. M. Leonard, D. W. Norbeck, H. Mo, and M. Markowitz. Mutations associated with the initial loss of suppression of plasma viral RNA levels in patients treated with ritonavir. *J. Virol.*, 72:5154–64, 1998.

[9] P. Essunger and A.S. Perelson. Modeling HIV infection of CD4+ T-cell subpopulations. *J. Theoret. Biol.*, 170:367–91, 1994.

[10] D. Finzi, M. Hermankova, T. Pierson, L.M. Carruth, C. Buck, R.E. Chaisson, T.C. Quinn, K. Chadwick, J. Margolick, R. Brookmeyer, J. Gallant, M. Markowitz, D.D. Ho, D.D. Richman, and R.F. Siliciano. Identification of a reservoir for HIV-1 in patients on highly active antiretroviral therapy. *Science*, 278:1295–1300, 1997.

[11] A.T. Haase, K. Henry, M. Zupancic, G. Sedgewick, R.A. Faust, H. Melroe, W. Cavert, K. Gebhard, K. Staskus, Z.Q. Zhang, P.J. Dailey, H.H. Balfour Jr, A. Erice, and A.S. Perelson. Quantitative image analysis of HIV-1 infection in lymphoid tissue. *Science*, 274:985–9, 1996.

[12] J.K. Hale and S.M. Verduyn Lunel. *Introduction to Functional Differential Equations.* Springer-Verlag, 1993.

[13] D.V. Havlir, S. Eastman, A. Gamst, and D.D. Richman. Nevirapine-resistant human immunodeficiency virus: kinetics of replication and estimated prevalence in untreated patients. *J. Virol.*, 70:7894–9, 1996.

[14] V.M. Herz, S. Bonhoeffer, R.M. Anderson, R.M. May, and M.A. Nowak. Viral dynamics *in vivo*: Limitations on estimations on intracellular delay and virus decay. *Proc. Natl. Acad. Sci. USA*, 93:7247–7251, 1996.

[15] D. D. Ho. Toward HIV eradication or remission: The tasks ahead. *Science*, 280:1866–1867, 1998.

[16] D.D. Ho, A.U. Neumann, A.S. Perelson, W. Chen, J.M. Leonard, and M. Markowitz. Rapid turnover of plasma virions and CD4 lymphocytes in HIV-1 infection. *Nature*, 373:123–6, 1995.

[17] T Hraba and J. Dolezal. Mathematical model of CD4+ lymphocyte depletion in HIV infection. *Folia Biol.*, 35:156–63, 1989.

[18] V. Isham and G. Medley, editors. *Models of Infectious human diseases.* Cambridge University Press, 1996.

[19] T.B. Kepler and A.S. Perelson. Drug concentration heterogeneity facilitates the evolution of drug resistance. *Proc. Natl. Acad. Sci. USA*, 95:11514–19, 1998.

[20] W.O. Kermack and A.G. McKendrick. A contribution to the mathematical theory of epidemics. *Proc. R. Soc. London*, 115:700–21, 1927.

[21] R.A. Koup, J.T. Safrit, Y. Cao, C. A. Andrews, G. Mcleod, W. Borkowsky, C Farthing, and D.D. Ho. Temporal assocation of cellular immune responses with the initial control of viremia in primary human immunodeficiency virus type 1 syndrome. *J. Virol.*, 68:4650–55, 1994.

[22] L.M. Mansky and H.M. Temin. Lower in vivo mutation rate of human immunodeficiency virus type 1 than that predicted from the fidelity of purified reverse transcriptase. *J. Virol.*, 69:5087–94, 1995.

[23] A. R. McLean and T. L. B. Kirkwood. A model of human immunodeficiency virus (HIV) infection in T-helper cell clones. *J. Theoret. Biol.*, 147:177–203, 1990.

[24] A.R. McLean and M.A. Nowak. Models of interactions between HIV and other pathogens. *J. Theoret. Biol.*, 155:69–86, 1992.

[25] J.E. Mittler, M. Markowitz, D.D. Ho, and A.S. Perelson. Refined estimates for HIV-1 clearance rate and intracellular delay. *AIDS*, 13:1415–17, 1999.
[26] J.E. Mittler, B. Sulzer, A.U. Neumann, and A.S. Perelson. Influence of delayed virus production on viral dynamics in HIV-1 infected patients. *Math. Biosci.*, 152:143–63, 1998.
[27] D. Mollison. *Epidemic models: Their structure and relation to data*. Cambridge University Press, 1995.
[28] P.W. Nelson, J. Mittler, and A.S. Perelson. Effect of drug efficacy and the eclipse phase of the viral life cycle on estimates of HIV-1 viral dynamic parameters. *AIDS*, page in press, 2001.
[29] P.W. Nelson, J.D. Murray, and A.S. Perelson. A model of HIV-1 pathogenesis that includes an intracellular delay. *Math. Biosci.*, 163:201–15, 2000.
[30] P.W. Nelson and A.S. Perelson. *Mathematical Analysis of Delay Differential Equations for HIV*. submitted, 2001.
[31] A.U. Neumann, N.P. Lam, H. Dahari, D.R. Gretch, T.E. Wiley, T.J. Layden, and A.S. Perelson. Hepatitis C viral dynamics in vivo and the antiviral efficacy of interferon-alpha therapy. *Science*, 282:103–7, 1998.
[32] M.A. Nowak, S. Bonhoeffer, G.M. Shaw, and R.M. May. Anti-viral drug treatment: Dynamics of resistance in free virus and infected cell populations. *J. Theoret. Biol.*, 184:205–19, 1997.
[33] M.A. Nowak, S. Bonhoffer, A.M. Hill, R. Boehme, H.C. Thomas, and H. McDade. Viral dynamics in hepatitis B virus infection. *Proc. Natl. Acad. Sci. USA*, 93:4398–402, 1996.
[34] M.A. Nowak and R.M. May. *Virus Dynamics: Mathematical Principles of Immunology and Virology*. Oxford University Press, 2000.
[35] A. S. Perelson. *Two theoretical problems in immunology: AIDS and epitopes. In Complexity: Metaphors, Models and Reality, G. Cowan, D. Pines and D. Meltzer, eds.* Addison-Wesley, Reading, CA, 1994.
[36] A.S. Perelson, P. Essunger, Y. Cao, M. Vesanen, A. Hurley, K. Saksela, M. Markowitz, and D.D. Ho. Decay characteristics of HIV-1-infected compartments during combination therapy. *Nature*, 387:188–91, 1997.
[37] A.S. Perelson, P. Essunger, and D.D. Ho. Dynamics of HIV-1 and CD4+ lymphocytes *in vivo*. *AIDS*, 11 (suppl A):S17–S24, 1997.
[38] A.S. Perelson, D.E. Kirschner, and R. De Boer. Dynamics of HIV infection of $CD4^+$ T cells. *Math. Biosci.*, 114:81–125, 1993.
[39] A.S. Perelson and P.W. Nelson. Mathematical models of HIV dynamics *in vivo*. *SIAM Review*, 41:3–44, 1999.
[40] A.S. Perelson, A.U. Neumann, M. Markowitz, J.M. Leonard, and D.D. Ho. HIV-1 dynamics *in Vivo*: Virion clearance rate, infected cell life-span, and viral generation time. *Science*, 271:1582–86, 1996.
[41] A.N. Phillips. Reduction of HIV concentration during acute infection: Independence from a specific immune response. *Science*, 271:497–499, 1996.
[42] G. Reibnegger, D. Fuchs, A. Hausen, E.R. Werner, M.P. Dierich, and H. Wachter. Theoretical implications of cellular immune reactions against helper lymphocytes infected by an immune system retrovirus. *Proc. Natl. Acad. Sci. USA*, 84:7270–4, 1987.
[43] D. Schenzle. A model for AIDS pathogenesis. *Stat. Med.*, 13:2067–79, 1994.
[44] R. Schuurman, M. Nijhuis, R. Van-Leeuwen, P. Schipper, D. De-Jong, P. Collis, S.A. Danner, J. Mulder, C. Loveday, and C. Christopherson. Rapid changes in human immunodeficiency virus type 1 RNA load and appearance of drug-resistant virus populations in persons treated with lamivudine (3TC). *J. Infect. Dis.*, 171:1411–9, 1995.
[45] M. Stafford, L. Corey, Y. Cao, E. Daar, D. Ho, and A. Perelson. Modeling plasma virus concentration during primary infection. *J. Theor. Biol.*, 203:285–301, 2000.
[46] N.I. Stilianakis, C.A.B. Boucher, M.D. DeJong, R. VanLeeuwen, R. Schuurman, and R.J. DeBoer. Clinical data sets on human immunodeficiency virus type 1 reverse transcriptase resistant mutants explained by a mathematical model. *J. Virol.*, 71:161–68, 1997.
[47] X. Wei, S.K. Ghosh, M.E. Taylor, V.A. Johnson, E.A. Emini, P. Deutsch, J.D. Lifson, S. Bonhoeffer, M.A. Nowak, B.H. Hahn, S.S. Saag, and G.M. Shaw. Viral dynamics in human immunodeficiency virus type 1 infection. *Nature*, 373:117–22, 1995.
[48] L.M. Wein, R.M. D'Amato, and A.S. Perelson. Mathematical considerations of antiretroviral therapy aimed at HIV-1 eradication or maintenance of low viral loads. *J. Theoret. Biol.*, 192:81–98, 1998.
[49] D.O. White and F.J. Fenner. *Medical Virology*. Academic Press, 1994.

[50] J.K. Wong, M. Hezareh, H.F. Gunthard, D.V. Havlir, C.C. Ignacio, C.A. Spina, and D.D. Richman. Recovery of replication-competent HIV despite prolonged suppression of plasma viremia. *Science*, 278:1291–95, 1997.

Theoretical Biology and Biophysics Group, Theoretical Division, Los Alamos National Laboratory, Los Alamos, NM 87545 U.S.A.
E-mail address: asp@lanl.gov

University of Michigan, Department of Mathematics, Ann Arbor, MI 48197 U.S.A.

Index

Titles in This Series

59 **James Sneyd, Editor,** An introduction to mathematical modeling in physiology, cell biology, and immunology (New Orleans, Louisiana, January 2001)

58 **Samuel J. Lomonaco, Jr., Editor,** Quantum computation: A grand mathematical challenge for the twenty-first century and the millennium (Washington, DC, January 2000)

57 **David C. Heath and Glen Swindle, Editors,** Introduction to mathematical finance (San Diego, California, January 1997)

56 **Jane Cronin and Robert E. O'Malley, Jr., Editors,** Analyzing multiscale phenomena using singular perturbation methods (Baltimore, Maryland, January 1998)

55 **Frederick Hoffman, Editor,** Mathematical aspects of artificial intelligence (Orlando, Florida, January 1996)

54 **Renato Spigler and Stephanos Venakides, Editors,** Recent advances in partial differential equations (Venice, Italy, June 1996)

53 **David A. Cox and Bernd Sturmfels, Editors,** Applications of computational algebraic geometry (San Diego, California, January 1997)

52 **V. Mandrekar and P. R. Masani, Editors,** Proceedings of the Norbert Wiener Centenary Congress, 1994 (East Lansing, Michigan, 1994)

51 **Louis H. Kauffman, Editor,** The interface of knots and physics (San Francisco, California, January 1995)

50 **Robert Calderbank, Editor,** Different aspects of coding theory (San Francisco, California, January 1995)

49 **Robert L. Devaney, Editor,** Complex dynamical systems: The mathematics behind the Mandlebrot and Julia sets (Cincinnati, Ohio, January 1994)

48 **Walter Gautschi, Editor,** Mathematics of Computation 1943–1993: A half century of computational mathematics (Vancouver, British Columbia, August 1993)

47 **Ingrid Daubechies, Editor,** Different perspectives on wavelets (San Antonio, Texas, January 1993)

46 **Stefan A. Burr, Editor,** The unreasonable effectiveness of number theory (Orono, Maine, August 1991)

45 **De Witt L. Sumners, Editor,** New scientific applications of geometry and topology (Baltimore, Maryland, January 1992)

44 **Béla Bollobás, Editor,** Probabilistic combinatorics and its applications (San Francisco, California, January 1991)

43 **Richard K. Guy, Editor,** Combinatorial games (Columbus, Ohio, August 1990)

42 **C. Pomerance, Editor,** Cryptology and computational number theory (Boulder, Colorado, August 1989)

41 **R. W. Brockett, Editor,** Robotics (Louisville, Kentucky, January 1990)

40 **Charles R. Johnson, Editor,** Matrix theory and applications (Phoenix, Arizona, January 1989)

39 **Robert L. Devaney and Linda Keen, Editors,** Chaos and fractals: The mathematics behind the computer graphics (Providence, Rhode Island, August 1988)

38 **Juris Hartmanis, Editor,** Computational complexity theory (Atlanta, Georgia, January 1988)

37 **Henry J. Landau, Editor,** Moments in mathematics (San Antonio, Texas, January 1987)

36 **Carl de Boor, Editor,** Approximation theory (New Orleans, Louisiana, January 1986)

35 **Harry H. Panjer, Editor,** Actuarial mathematics (Laramie, Wyoming, August 1985)

34 **Michael Anshel and William Gewirtz, Editors,** Mathematics of information processing (Louisville, Kentucky, January 1984)

33 **H. Peyton Young, Editor,** Fair allocation (Anaheim, California, January 1985)

32 **R. W. McKelvey, Editor,** Environmental and natural resource mathematics (Eugene, Oregon, August 1984)

31 **B. Gopinath, Editor,** Computer communications (Denver, Colorado, January 1983)
30 **Simon A. Levin, Editor,** Population biology (Albany, New York, August 1983)
29 **R. A. DeMillo, G. I. Davida, D. P. Dobkin, M. A. Harrison, and R. J. Lipton,** Applied cryptology, cryptographic protocols, and computer security models (San Francisco, California, January 1981)
28 **R. Gnanadesikan, Editor,** Statistical data analysis (Toronto, Ontario, August 1982)
27 **L. A. Shepp, Editor,** Computed tomography (Cincinnati, Ohio, January 1982)
26 **S. A. Burr, Editor,** The mathematics of networks (Pittsburgh, Pennsylvania, August 1981)
25 **S. I. Gass, Editor,** Operations research: mathematics and models (Duluth, Minnesota, August 1979)
24 **W. F. Lucas, Editor,** Game theory and its applications (Biloxi, Mississippi, January 1979)
23 **R. V. Hogg, Editor,** Modern statistics: Methods and applications (San Antonio, Texas, January 1980)
22 **G. H. Golub and J. Oliger, Editors,** Numerical analysis (Atlanta, Georgia, January 1978)
21 **P. D. Lax, Editor,** Mathematical aspects of production and distribution of energy (San Antonio, Texas, January 1976)
20 **J. P. LaSalle, Editor,** The influence of computing on mathematical research and education (University of Montana, August 1973)
19 **J. T. Schwartz, Editor,** Mathematical aspects of computer science (New York City, April 1966)
18 **H. Grad, Editor,** Magneto-fluid and plasma dynamics (New York City, April 1965)
17 **R. Finn, Editor,** Applications of nonlinear partial differential equations in mathematical physics (New York City, April 1964)
16 **R. Bellman, Editor,** Stochastic processes in mathematical physics and engineering (New York City, April 1963)
15 **N. C. Metropolis, A. H. Taub, J. Todd, and C. B. Tompkins, Editors,** Experimental arithmetic, high speed computing, and mathematics (Atlantic City and Chicago, April 1962)
14 **R. Bellman, Editor,** Mathematical problems in the biological sciences (New York City, April 1961)
13 **R. Bellman, G. Birkhoff, and C. C. Lin, Editors,** Hydrodynamic instability (New York City, April 1960)
12 **R. Jakobson, Editor,** Structure of language and its mathematical aspects (New York City, April 1960)
11 **G. Birkhoff and E. P. Wigner, Editors,** Nuclear reactor theory (New York City, April 1959)
10 **R. Bellman and M. Hall, Jr., Editors,** Combinatorial analysis (New York University, April 1957)
9 **G. Birkhoff and R. E. Langer, Editors,** Orbit theory (Columbia University, April 1958)
8 **L. M. Graves, Editor,** Calculus of variations and its applications (University of Chicago, April 1956)
7 **L. A. MacColl, Editor,** Applied probability (Polytechnic Institute of Brooklyn, April 1955)

For a complete list of titles in this series, visit the AMS Bookstore at **www.ams.org/bookstore/**.